Broken Trust, Broken Land

Freeing Ourselves From The War Over The Environment

Robert G. Lee

BookPartners
Wilsonville, Oregon

BookPartners, Inc.
P.O. Box 922
Wilsonville, Oregon 97070

Table Of Contents

To my father, John,
and mother, Katharine,
whose lives inspired this book.

Preface

Some have observed that the life of a professor in a modern research university is dedicated to knowing more and more about less and less. We know more about particular things—how attitudes are formed, how individuals are affected by loss of a job, how enzymes work, how redwood trees pump water 300 feet above ground level. This is important work, but in doing it we tend to lose touch with the realities about how all these things are connected and what they ultimately mean. We lose touch with the larger mission of seeking truth and teaching truth.

This book is a search for truths about how Americans are losing themselves in their attempt to solve environmental problems. It doesn't summarize research findings or review theories, and should not be read as a report on findings, or even as a conventional academic treatise. Hence, this book stands somewhere between prophecy and scholarship, and was written for non-technical readers as well as for other natural resource professionals and social scientists. For this reason, I have not burdened the reader with voluminous citations and technical details, and limited attribution to sources for key ideas.

I argue that many Americans are exchanging their freedom for environmental security, and need to think about whether they really want to take such a momentous step. Since this book is about freedom, it is neither neutral nor "value free." Liberty and truth are about values, so my remarks can't be free from values. Truth has to do with who we are, why we are here, how we should live our lives, and how we should treat other people and our surroundings.

This book is also about telling truths. Telling lies and half-truths has become acceptable as contestants jockey for position in the war over the environment. Propaganda is deliberately used by all sides in this battle. Scientists have found their work distorted by political agendas or have advocated solutions by telling only part of what they know to be true. Even when not advocating a position, most of us hold back and tell only part of what we think is true.

This book violates a solemn rule I learned from working in the modern research university: truth is like a powerful healing drug and should be dispensed only in small doses for tightly circumscribed problems. I have taken the risk of violating this rule, pulling out all stops, and telling it like I see it. Readers who find this discomforting should plan on pacing their reading in small steps so that my search for truths can be absorbed without unintended reactions. My purpose is to stimulate reflective thought, not anger, fear, or a hopeless state of confusion.

We live in a rapidly changing world. We can most effectively adapt to this change if we have a stable reference point that provides us with a clear sense of identity. Managers of modern organizations understand that successful change is possible if we maintain independence from the environment. Margaret Wheatley captured this when she said that

"... a clear sense of identity — of values, traditions, aspirations, competencies, and culture that guide the operation — is the real source of independence from the environment. When the environment demands a new response there is a reference point for change. This prevents the vacillations and the random search for ... new ventures that have destroyed so many businesses over the past several years. [1]

An important role for a scholar is to provide both students and citizens with an appreciation for how the development and maintenance of personal and cultural identities gives people the reference points needed for evaluating change and making choices. Without such reference points, people are likely to fall into the sorts of personal confusion and social chaos that have destroyed individuals and societies. We all have our points of view — or reference points. This book expresses my reference point.

Freedom is fundamental to who I am and what I believe. I thrive on both political and spiritual freedom. My experience as a rural American taught me that freedom and community were not opposites, as is too often believed among modern urban academics and government administrators. I have been encouraged by academics who understand how orderly social life can arise from freedom. Friedrich Hayek devoted much of his life to showing how economic freedom contributes to an orderly society. And modern scientists are discovering how order in nature and society arises from freedom and the fluctuations it brings.

Any serious search for truths is a personal story. Hence, this book is, in large part, a chronicle of my personal experiences as a forester who acquired sociological expertise in an attempt to address complex problems of environmental management. Most importantly, I do not speak primarily as a sociologist or a forester. In this book I instead speak as a concerned American citizen who is searching for truths that transcend the particular issues facing scientists. I talk about science as a useful means for seeking truths because it offers the discipline we need to keep from deceiving ourselves or deceiving others. But I also recognize other useful ways for seeking truths, including artistic expression, love, meditation, prayer, and debate.

I have traveled a full circle and returned to where I began. And, as T. S. Eliot said in his poem, "Little Gidding," "I have returned to the place I started, and know it for the first time." The pursuit of political and spiritual freedom has been a constant motive in this journey. I learned the value of freedom early in my life while growing up on ranch and living in a rural community. Yet freedom eluded me when I worked summers as a Forest Service timber management trainee In the Mendocino National Forest. Experiences with "government work" led me to choose a career in the forest products industry. I worked as a forester for a small company in the Redwood Region of northern California, but was soon disappointed by external corporate control that degraded both forest land and the people in rural communities who worked this land.

The idealism of the 1960s inspired me to study social philosophy and then sociology in an attempt to promote better land management and human welfare. I was drawn by the optimistic dream that people could be educated to treat others with more respect and do a better job of managing land. I combined a firm belief in the potential for dramatic human improvement with an ecosystem approach to land management. I pursued this dream for over 15 years before these idealistic beliefs were shaken by the realization that much that I had worked for was a far greater threat to freedom than that I had sought to change.

I discovered that those with whom I had worked to promote human improvement and ecosystem management were far more interested in controlling other people than they were in giving people meaningful control over their lives. Freedom had become an archaic sentiment, and even students began to look upon me as naive for thinking that freedom was essential for conserving forests, wildlife, water, land and other resources. Academic research and teaching increasingly places decision-making authority in the hands of experts whose "understanding" is seen as sufficient for deciding both how land should be managed and how the people who work it should live. My dreams of freedom have been replaced by a nightmare of growing control by increasingly politicized academic and scientific elites.

My search for freedom has returned to where it began. I have returned to people who manage land and produce necessary things from it. I have returned to the conservative wisdom of traditional American institutions as the best means for protecting freedom and community. And, along with much of the world community, I have discovered that maintenance of freedom is the best means for sustaining production from land while protecting the environment. I have reaffirmed the value of America's resilient political institutions to protect us from the unintended consequences of those who want to do good but fail to understand how good actually gets done.

Some readers may be troubled that I have taken a point of view based on my personal experiences and beliefs. Please understand that I am speaking as a concerned citizen, not simply as a scientist, when I criticize others for politicizing science or suggest how freedom can be restored to the American political landscape. I reveal my own values and beliefs because I want to show how other values and beliefs are embedded in much that passes as ecology, environmental science, or social science. I also want my readers to understand that conflicts over the environment are primarily moral and political issues, and are only secondarily matters of science. Hence, these issues cannot be resolved by science or science-based plans. We instead urgently need active political debate centered on the moral choices that face us.

I am convinced we will resort to fighting one another unless we start talking about these choices. American citizens will not know that they have a *choice* over values and beliefs unless they can see these contrasts. I draw on years of experience and study to set forth principles on how we can best produce commodities from the environment while protecting its essential ecological functions. And because rural producers hunger for something constructive to do during this period of rapid change, I have suggested ways in which these principles could be implemented. But, ultimately, my purpose is to argue for restoring and protecting freedoms that have been undermined by the crusade to "Save the Earth." I am convinced that we cannot succeed in protecting the environment unless we first protect freedom.

Acknowledgments

None of the ideas in this book is original. I can only claim responsibility for how I have pulled them together and focused them on our attempts to live with one another and with the land that provides our material needs. Several scholars deserve recognition for the profound impact they have had on my thinking. Henry Vaux is my primary mentor in forestry, and gave me the courage to be myself and resist social control by governing elites, professional zealots, and mass movements. Walter Firey and William Burch introduced me to the sociological study of natural resources. Herbert Blumer patiently taught me the symbolic interactionist method. Charles Lindblom, both as author and instructor, taught me the value of democracy. Thomas Sowell, Russell Kirk, and Frederick Hayek led me to see the essential contributions of conservative principles.

Rene Girard helped me strip off the superficial layers of modern religion and rediscover the radical message of Christ. Tikva Frymer-Kensky led me to reaffirm monotheism by allowing me to look at religion through the eyes of an outstanding feminist scholar. Stephan L. Carter clarified my views on the need for religious autonomy. George Thomas, Norman Cohn, Serge Moscovici, Eric Hoffer, and James McEvoy led me to a window where I could look at the environmental movement as a monumental force in cultural change. Zygmunt Bauman and Richard Rubenstein helped me see the seeds of tyranny and oppression in modern rationality and science. Robert J. Lifton and P. W. Martin took me deep into myself and showed me how to own what I had long refused to face. But I have been inspired most by rural Americans who steadfastly believe in the rule of law and have faith in their nation even as their government violates its own laws and betrays the moral foundations upon which the nation was founded.

I want to thank Alton Chase, John Baden, and Carla Berkedal for their helpful review and comments on an early version of this manuscript. Bob Legg, President of the Temperate Forest Foundation, has inspired many of the ideas I express in the third section of this book. My publishers, Thorn and Ursula Bacon, deserve special thanks both for their patience and faith in my work and for Ursula's creative design of the cover. But forthright students have been my best critics, and have taught me more than I have taught them. I am especially indebted to Penny Eckert and Kristin Warren for their critical reading of this manuscript. The opinions expressed here are my own, and do not necessarily represent the views of my reviewers.

But most thanks should go to my wife, Karen—a constant friend and companion in a world of turbulence, deception, and betrayals. She continues to embody the enduring truths of faith and friendship.

Introduction

The Game Is "War"

We must fight 'self'.

Mao Tse-Tung

To secure ourselves against defeat lies in our own hands, but the opportunity of defeating the enemy is provided by the enemy himself. Hence the saying: One may know how to conquer without being able to do it.

Sun Tzu
The Art of War

1

At War
With
Ourselves

"Education" tries to create [people] who will "autonomously" serve col-lective interests, that is, who will do on their own initiative what in other societies they must be commanded or induced to do. It must also create [people] who will voluntarily respond to state and party when either asks for specific performance.

Charles Lindblom
Politics and Markets

The memories of his life as a logger hung from the soft green walls of the living room where I sat talking with his wife: "cork" (hob-nailed) boots, a crudely painted portrait of himself standing proudly beside his first log-loader, photographs and awards from logging shows, and a worn hickory shirt draped over the back of his easy-chair. The walls spoke of the pain of a proud man stunned by clever words he could not understand.

"He doesn't like to talk much," she continued, as I gazed about me at the remnants of his life. "But he sure loves those woods, and he'll talk about them all day long if you let him. He'll be comin' home from *his turn at the food bank soon, and if you'd ask him, he'll take you to the woods and show you around."*

The battered Ford pick-up climbed over a ridge, crossed into the national forest, and came to a stop on an old landing covered by young trees. "We finished this unit in the fall of '81—got the iron [equipment] out just before the snow hit. Just look at this. These trees are jumpin' out of the ground so fast you'd think God himself had meant this to be."

We got out of the pick-up, put on our rain slickers, and silently pushed our way under young trees until we reached a small lake bordered on the other three sides by tall trees. "I've been comin' here to fish since I was five. My father used to bring me on Saturday afternoons. So you can see why I was damn careful to keep the brush out of here when we cut this unit. And you know what, the fishin's been better than ever the last five years. My son, Bill, limited here last weekend, and you should've seen that boy grin when he laid those fish on the sink for his ma to fry. First meat we'd had in a week-and-a-half!"

His steel blue Swedish eyes lost their twinkle and turned gray as his smile vanished into an empty gaze. He silently searched the lake for almost five minutes. I couldn't tell what he was thinking, and began to feel uneasy.

"Too damn bad! I don't know what the world's comin' to these days. Never seen anything like it in all my life. I've worked hard all my life, fought a war for this country, paid my taxes, supported my wife and kids, stayed out of people's way … I … you …," he paused, and probed my eyes with a question suspended on waves of emotion, "Mary says you're a sociologist and understand these things. Now you tell me, Professor, what the hell is happening to this country? Sure, we made our mistakes like everyone else in life, but we did the best we could, and we have served our country well. Why are good, hard-workin' people tossed out like so much garbage and left to rot? What did we do to deserve this?"

He turned away, and searched the sky for the tips of trees hidden in the low clouds. Even if his gaze had continued to bore through me, I could not have seen the tears in his eyes. I walked over to the edge of the pond, took out my handkerchief, and pretended to blow my nose as I dried my eyes. I was tortured by his pain and ashamed that I couldn't answer his questions. I knew it and he knew it. He probably knew it before he asked. And he knew it before I did. "Damned academic derelicts," I could hear him thinking, "Know 'bout everything but things that matter most."

I met more people and was asked more questions. I met Sam, a contractor who worked hard, saved for years, bought a piece of property in the country on which to build his dream house, and then found that wetlands regulations prohibited all but minimal new construction. He couldn't recover his savings and build elsewhere because all he worked for was tied up in land that he could not use or sell for what it was worth before the regulations were issued. The public at large would benefit, but nobody compensated Sam for what was taken from him.

I met Mary, who inherited her grandfather's ranch and wanted in the worst way to protect the beautiful forests and meadows that had been a joy to her family for three generations. But sheep and cattle ranching wasn't as profitable as it used to be, and she wasn't able to pay for maintenance and property taxes without selectively cutting some of the trees every few years. Then she learned that the land was within a designated habitat conservation area for an endangered species and that no timber harvesting might be allowed after new regulations took effect in six months. She was faced with the awful choice of possibly losing the ranch or immediately cutting all the merchantable timber and transferring the receipts to a safer investment. It broke her heart when she liquidated the timber assets and left the land to recover in hopes that her great-grandchildren could someday enjoy what she had known as a child.

I met Orville and Mable, who feared they would lose their ranch if the federal government imposed high fees on grazing permits or entirely eliminated grazing on public land. They were confused and angry because the government's own records showed that range conditions had improved steadily over the four generations their family has managed these lands. They were only too willing to make the adjustments necessary to further improve range conditions and protect the streams because they wanted to maintain their way of life and extend their unbroken chain of land stewardship to their daughter and her children.

Their questions were much the same: What is happening to us? Why don't people see how they are making us suffer? Why won't they accept the facts that the nation's forests are growing faster than they are being cut? Why won't anyone believe that the rangelands are in better shape than they have been in 100 years? Why do people so easily believe the lie that we are working our way out of a job by destroying the last of the forests? Why can't they see the vast areas that have been preserved and our willingness to set aside the best of the remaining natural amenities? Why don't they see how we are helping the rest of the country by producing what people want? How could this happen in America? Why won't people talk about what is happening? What has happened to people that makes them so uncaring, so cruel? What can we do make things right again? The burden of their unanswered questions weighed on me, and I began to think and read, and ask these questions of myself. And I began to write this book.

Writing is indeed an act of thinking. When I completed the first draft and read the manuscript through for the first time, I was startled to discover I could now confidently look people in the eye and answer all the questions they had asked. Scattered pieces fell together, and the picture was complete. I had discovered how the "rules of the game" are being changed. I found that change is so rapid that the game itself may be changing, and those who are changing the rules of the game are not telling others what the new game is or how it is supposed to be played. Not talking about the new rules gives those pushing change a big advantage. Playing by the old rules in a new game robs people of their power and makes them vulnerable to manipulation. Just as Sun Tzu had written 2,500 years ago, people provide the opportunity for their own defeat by failing to understand the rules of the game into which they are drawn.

Without knowing it, I had actually been writing about the emergence of "moral persuasion" as a dominant means of social control. I had discovered the new game and am now in the process of deciphering its rules. This book answers all the questions I was asked by describing the game of moral persuasion, revealing its hidden dangers, and explaining why restoration of traditional American institutions offers a far better way for us to sustain our society and environment.

What do I mean by "moral persuasion"? Why has it been so

characteristic of the environmental movement? Why is it so effective in smothering the free expression of ideas? And why did it so easily take hold in the United States? These questions will be examined in this chapter and the two that follow. I will talk about what has happened to make people disregard human suffering in rural America.

The questions asked by rural people forced me to look at myself. The second section of the book will describe how I discovered how I too had fallen victim to moral persuasion and had to struggle to free myself from the invisible controls it had implanted in the dark reaches of my self. I had given up too much of myself to academic absolutism and had not studied the things that mattered most.

I recovered my freedom by deciphering the game of moral persuasion, and believe others can do the same. Hence, the last five chapters explain how rural people can empower themselves and help lead the nation in restoring freedom to the American political landscape.

Getting The Masses To Do What Elites Want

Moral persuasion is a systematic way of controlling an entire population by "teaching" them that one way of thinking or acting is better than others.[2] A dominant elite—a powerful leadership group respected for its superior knowledge or insight—tries to get people to think and act in a uniform way so that centrally-desired aspirations can be realized. Moral persuasion is not "top-down" implementation of centralized goals through a rigid bureaucracy. And neither is it "bottom-up" management by empowering citizens to choose goals and develop programs for implementing them. It is both "top-down" and "bottom-up," since it enlists citizens in a process of voluntarily implementing the centralized goals.

Moral persuasion has been used effectively by an elite consisting of respected scientists, environmental advocates, politicians, teachers, journalists, and clergy. I will use the terms "environmentalists," or "environmental preservationists," to refer to this loosely connected group of leaders. Although the environmental movement is exceptionally diverse, and contains all kinds of people with differing points of view, committed environmentalists are united by a belief in an impending crisis that necessitates personal transformation and commitment to "saving the earth." A shared climate of fear, *not* conspiracy, has activated influential leaders.

My purpose is to show how this elite is succeeding by using rules that differ from what people have used in the past. Rural people had learned how to work with traditional local elites, including, but not limited to, the local and regional press, elected officials, educators, labor unions, and business and industrial leaders in agriculture, ranching, forest products, and mining. They gained partial control over their lives by playing the game of local and regional democratic politics. But they were totally unprepared for a national elite that swamped traditional elites by playing a new game with unwritten rules.

Charles Lindblom, a Yale economist and political scientist, has compared various forms of social control in governments throughout the world. He characterized moral persuasion as a "preceptoral system" of social control. "Preceptor" means teacher. The "teacher" is generally a governmental or social elite which assumes the task of instructing masses of people. The masses of people are assumed to be blank slates on which the "teacher" writes the "superior intelligence and understanding" of those who are "enlightened." "Education" comes closer to describing what moral persuasion does, although indoctrination, instruction, propaganda, counseling, advice, exhortation, and thought control are all involved.

Optimistic Dreams Of Virtuous People
Plato's *Republic* is the first known example of someone advocating moral persuasion. Plato thought "philosopher-kings" (what some have referred to as "benevolent dictators") should rule and instill virtuous behavior in the masses. His vision of a utopian society was based on an untested assumption that has haunted all utopian experiments for the last two thousand years. Plato had an optimistic view of people's willingness and ability to produce good. People were seen as having an untapped potential for virtuous behavior. All that was needed to realize good was the centralization of power in the hands of enlightened leaders who would provide the proper instruction and free citizens from constraints imposed by the oppressive forces of ignorance, greed, and social prejudice.

American Pessimism Built Democracy

American government was founded on a pessimistic assumption about the human potential for producing good. Our forefathers established a government based on law, not the good intentions of people.[3] They carefully designed a system of checks and balances, involving the executive, legislative, and judicial branches of government. Experience had taught them that individuals could not always be trusted to do what was right for the country, and sometimes not even what was right for themselves. Rules were made predictable and secure by embedding them in law. Free speech, freedom of religion, freedom of assembly, private property, and an open market economy all were seen as means for protecting citizens from falling under the control of a powerful ruler or elite group.

Creating The "New Person"

Moral persuasion rejects the conservative wisdom of our founding fathers and starts from the optimistic assumption that humans are capable of spontaneous, virtuous behavior. Above all else, "education" seeks to *transform* the personality and to create a "new person." Many of the New Age beliefs and practices reflect a faith in the transformational power of "education." The human potential movement of the 1960s and 1970s was alive with such minor forms of moral persuasion. The "Age of Aquarius" anticipated by the 1960s "flower children" was expected to bring massive transformation in people. Many of the religious cults that sprang up during this period also reflected this same emphasis on transformation. Modern advertising resembles moral persuasion in its attempt to sway people into thinking one product is superior to another, but its effectiveness is blunted by open competition among advertisers. And many modern management methods emphasizing a "democratic workplace" embody the faith that average workers can be transformed into creative problem-solvers who will "spontaneously" implement the centralized goals of a profit-making enterprise.

Fighting The 'Self' In China And Cuba

Lindblom tells us that the potential for moral persuasion has never been fully tapped, but that it was most highly developed by communist governments in China and Cuba: "Mao speaks of the need to 'remold people to their very souls.' 'We must fight "self".' Castro declared that the 'fundamental task is the formation of the new man, a man with profound consciousness of his role in society and his duties and social responsibilities.'"[4] Extreme selflessness, cooperation, egalitarianism, and service to society were Mao's and Castro's images of the "new person," along with duty, hard work, and self-discipline.

The fight against "self" is not designed to defeat or destroy individuals, as is typical of war or of the sort of coercive administrative hierarchies found in Nazi Germany or Russia under the Soviet Union. The purpose of the "fight" or "struggle" is to stun and immobilize the self so that it may be colonized and transformed. Social control is not secure until "ownership" of selves falls into the hands of those implementing centralized objectives. Hence, this book revolves around three questions: *Who owns our selves? How is ownership of our selves transferred to the elite? How can we protect our selves from being controlled by an influential few who advocate radical environmental reforms?*

I contrast the "teacher's" attempt to colonize and transform the self with the original, root meaning of education, which was to "open the self"—to help individuals gain greater awareness of who they are and greater responsibility for living. To "own" one's self is to take responsibility for living and to hold one's self accountable to higher principles or beliefs, often including a belief in God. This requires that we have firm values and clear principles to both guide us and enable us to know when we have met our responsibilities.

Eliciting Voluntary Conformity

Chinese and Cuban leaders were trying to achieve something that many American government, business, and environmental leaders have also sought: voluntary conformity of citizens to centrally directed initiatives. The goal was to turn the average person into a problem-solver who would spontaneously seek solutions to problems defined by those in control of government, profit-making, or "saving the Earth." But unlike American leaders, communist rulers backed up their use of moral persuasion with physical force. Rifles and work camps were generally sufficient to convince most people that it was wise for them to "voluntarily" conform to centralized directions.

Persuasion Is Not "Command And Control"

The combination of moral persuasion with the authority of a centralized bureaucracy has deluded many into thinking that readiness to use violence was the basis of social control in China and Cuba. Yet moral persuasion is unlike the sort of "command and control" structure and propensity for violence found in totalitarian states that rely on force. We fail to see China's primary means of social control when we explain conformity by attributing it to authoritative control by a centralized bureaucracy. China developed moral persuasion to a point where moral incentives were sufficient to control most people because citizens would voluntarily exercise these social controls by policing one another.

But, most importantly, we are also seriously misinformed when we explain environmental "correctness" in American society in terms of a reliance on "command and control." The U.S. government is increasingly relying on centralized authority and exercise of police powers to "correct" environmental problems, but it can only do so because a sufficient number of citizens believe this is the right thing to do. American citizens are becoming more puritanical in how they hold one another accountable to "correct" ways of thinking, speaking, and acting.

Institutionalized Conflict

American leaders have long realized what effective rulers have always known. Moral persuasion works best in times of national emergency, especially national mobilization for war. Average individuals will voluntarily become problem-solvers in an effort to meet the centralized objective of defending the nation by winning the war. All the Chinese and Cubans did was to make war a daily event. Chairman Mao effectively institutionalized the Communist Revolution by instigating continual purges and political cleansing rituals, the most turbulent of which was the Cultural Revolution. American leaders have had the same impulses, but, until recently, lacked the social instruments for creating the mass mentality necessary for perpetuating "war" mobilization.

Moral Equivalent of War

In 1906, the pacifist Harvard psychologist William James urged national leaders to create a "moral equivalent of war"[5] in order to direct aggressive passions away from war and toward more constructive purposes. Woodrow Wilson was the first leader to mobilize the economy and society for a foreign war, and the nation has subsequently been united by a war mobilization mentality.[6] Franklin Roosevelt effectively used such moral persuasion in both the depth of the Great Depression and in World War II. Then came the Cold War and the fight against Communists, both internal and external. Jimmy Carter abortively sought to create a "moral equivalent of war" in order to mobilize the nation to combat the 1970s energy crisis.

But no previous peacetime mobilization rivals the use of moral exhortation to mobilize people for "saving the Earth." For decades, conservation leaders have attempted to mobilize people to address environmental problems by inventing a sense of crisis—a "moral equivalent of war." This "Chicken Little" theory has been a singular strategy for motivating people to conserve forests, soils, and wilderness, eliminate pollution and waste, and reduce consumption of energy and materials. The battle to eliminate wildfire symbolized by Smokey the Bear was highly successful in persuading people that all fire was bad. The foolishness of this campaign left us with sick forests, crowded with unnatural accumulations of undergrowth and trees that explode in unnaturally intense fires. Now the fear of impending chaos once fixed on fire has been transferred to cutting trees, grazing cows, and building roads, homes, and businesses.

Freedom Is The Price Of Perpetual War

An ever-escalating threat cycle has been created. Fear of impending disaster attracts voluntary contributions which are used to fund "war-time" propaganda that spreads fear of new threats. This "conflict industry" is self-perpetuating. Legislative and judicial reforms are assumed to follow growing popular fears stimulated by escalating threats. Just as was the case in trying to rid forests of wildfires, the use of moral exhortation to induce fear has unanticipated consequences.

But, unlike mobilization against fire, the crusade to " Save the Earth" creates a mass mentality that can undermine the institutional basis for our democratic political system. As a result, we may increasingly resemble the collectivist societies we have successfully opposed for the last 60 years. We cannot be continuously mobilized for fighting a war and at the same time perpetuate freedom of thought, private property, and freedom of religion.[7] A war to " Save the Earth" necessarily centralizes control and erodes freedom. Leaders in the environmental movement, and the politicians and press who follow them, fail to understand the enormous political costs of centralized control focused on a singular objective.

A New 'Organizing Principle' For Civilization

Consider for a moment this quote from Vice-President Albert Gore's book, *Earth in the Balance: Ecology and the Human Spirit,* written when he was a U.S. Senator:

"...we now face the prospect of a kind of global civil war between those who refuse to consider the consequences of civilization's relentless advance and those who refuse to be silent partners in the destruction. More and more people of conscience are joining the effort to resist, but the time has come to make this struggle the central organizing principle of world civilization.[8]"

This is truly moral persuasion. A "struggle," a "civil war" to save the environment, not liberty, would become the "organizing principle" for all nations.

Not even Mao exceeded the compelling logic of internal and external struggle portrayed by Gore. Yet Gore clearly expressed his dedication to democracy and rejected any form of totalitarian government. He would be shocked to find himself in company with Mao. Gore and Mao are only alike in their understanding of the potential for moral exhortation to create the mass mentality necessary for radical change. They advocate the same method for advancing different forms of government. Although many of us find it disturbing, moral persuasion is not limited to communism or "foreign influences," but is instead a universal form of social control found to a certain extent in all societies and associated with all governments. Moral persuasion has, in one form or another, been with us since colonial times and is a very American trait.

Igniting a War in Our Selves

This book examines how our propensity for scientific rationalities and Calvinist beliefs predisposes us to the growth of moral persuasion—my topics for Chapters 2 and 3, respectively. Gore, like most other advocates for radical environmental reforms, was true to rationality and Calvinism when he authored *Earth in the Balance*. But he was also promoting a kind of conversion experience known to centralized religions, Mao, and others as well. Moral persuasion seeks to create, *or exploit*, a split in the individual's sense of self—an internal struggle between "good" and "evil" parts of the self—and then to colonize the self by implanting an idealized, or romanticized, "good," "new self" that would triumph over the "bad," "old self" and give individuals the illusion that they had become as "virtuous" as their "leaders."

Lindblom stated it this way:

Mao sees it as a continuing struggle within each individual, between an older and corrupt man and the new man. That being so, it calls for 'education' more than new forms of authority to destroy the old class.[9]

Gore was not concerned with overcoming the capitalist class, but with overcoming a "dysfunctional civilization" when he said:

The struggle to save the global environment is in one way much more difficult than the struggle to vanquish Hitler, for this time the war is with ourselves. We are the enemy, just as we have only ourselves as allies. ...It is not merely in the service of analogy that I have referred so often to the struggles against Nazi and communist totalitarianism, because I believe that the emerging effort to save the environment is a continuation of these struggles.[10]

Internal War Relinquishes Ownership Of Self

Moral persuasion seeks to promote an intense internal conflict which results in a transformation of the self. The old self is "bad" (not OK) and is to be replaced a new virtuous self (a "good self") inspired by the "teachers." This was the genius of the Cultural Revolution when people were forced to recant their old lives and adopt "correct" ways of thinking and living. And this is also the perverse genius of federally-sponsored "ecosystem restoration" programs that ask unemployed loggers to recant the destruction they caused in their former lives as loggers and adopt new lives by "repairing the damage they have done." Such transformations make the masses good problem-solvers for the "correct" regime of the "teachers."

Transplanting The "Ecological Self"

"Engineers of the soul" (Mao's term) have crafted a "new self" and seek every opportunity to teach its virtues. Listen to Alan Drengson, a "deep ecologist" whose comments help introduce the Sierra Club book, *Clearcut: The Tragedy of Industrial Forestry:*

> We have cleverly manufactured human abstractions so as to return to our whole, original, spontaneous experience in which there are no fixed boundaries between self and others. Professor Arne Naess, who coined the term *deep, long-range ecology movement*, uses the term *ecological self* to refer to broad and deep identification beyond our ego and physical bodies. In the broadest identification, our *ecological self* is intertwined with all beings. We can explore this *ecological self* by means of expanding our personal commitment. In doing so we enlarge our capacity for sharing and caring. We can then explore our many interconnections, if only through breathing. [original emphasis][11]

The campaign to stop clearcutting on all private and public lands condemns "old selves" by likening clearcutting to the Holocaust—even to the point of using a slide-show to display the *Clearcut* book in front of a mural of starving prisoners in a German concentration camp. The masses are invited to reject this "holocaust in the forest" and be reborn by committing themselves to mystical participation in undisturbed forests.

Identifying With The Group

The new self is fully prepared to "voluntarily" implement the centralized goals of the social, political, or business elite because it has been re-formed in the image of a mass identity—the elite's image of the virtuous person. Individuals acquire a "mass" or "group" identity when they come to see themselves primarily and exclusively (totally) as "Reds," "Greens," liberals, conservatives, gays, straights, women, men, blacks, whites, scientists, or corporate citizens—almost as followers of some religious cult. Their choices are governed by rigid scripts telling them what to say and how to live a "correct" life. These choices are not their own. The self is now "owned" by the cult: the state, the party, the corporation, the lifestyle, or powerful environmentalists.

Blunting Competing Moral Authorities

Moral persuasion has worked best when elites controlled communications and eliminated as many competing moral authorities as possible. State control over the mass media in totalitarian regimes brings communications under the control of ruling elites. But these regimes also limit the effectiveness of competing moral authorities by severely discouraging pluralistic associations and attachments. The state deliberately weakens the family, union, church, temple, mosque, community, employer, landlord, and other associations so that people will become more responsive to "education" at the hands of the state. This is why China and Cuba could be so successful in turning the state into the "teacher."

Voluntary Censorship

But China and Cuba never succeeded in achieving "voluntary" control over communications. How ironic that Americans would adopt forms of "correctness" that the Chinese and Cuban communists could not accomplish through coercion. And how doubly ironic that such moral censorship should emerge "spontaneously" in institutions founded on the principles of free expression—the universities and the mass media. How did we get to a place where to describe the plight of displaced loggers, ranchers, or farmers is to invite moral condemnation? How could this happen in America?

Excessive Individualism Can Kill Democracy

Alexis de Tocqueville, the French scholar who studied the American democracy 150 years ago, anticipated these changes.[12] He predicted that the drift of democracy toward excessive individualism would erode the foundations of our democratic institutions and cause people to embrace totalitarianism. Robert Bella and his associates said it this way:

> When economics is the main model of our common life, we are more and more tempted to put ourselves in the hands of the manager and the expert. If society is shattered into as many special interests as there are individuals, then, as Tocqueville foresaw, there is only the schoolmaster state to take care of us and keep us from one another's throat.[13]

De Tocqueville based his predictions on evidence of a growing weakness in social institutions that stand between individuals and the nation state or other large organizations that attempt to control the lives of individual citizens. These protective institutions have been named "mediating institutions" by sociologists because their more immediate social arrangements give individuals greater control over the events that affect their lives. The mediating institutions he talked about then are the same ones we talk about today when we worry about how to stop street violence. These are typically families, neighborhoods, local communities, local religious institutions, voluntary associations of all kinds, and a plurality of interest groups to which people make enormous personal and financial commitments. These mediating institutions are the vehicles through which Americans have best contributed to common purposes and collective goals.

Family guidance, religious teachings learned in small groups of believers, and democratic principles practiced in self-governing voluntary associations are all moral authorities. When individuals abandon these authorities in an attempt to maximize their freedom, they often succumb to the state, television messages, and a variety of mass movements as substitute moral authorities. I will talk more about the vain attempt to escape moral authority in Chapter 8.

Television Substitutes For Moral Authority

We should not underestimate the power of television as a replacement for the values previously taught by small gatherings of committed believers, strong families, or voluntary associations. Nothing in modern society is as effective in creating a "mass mentality" than the uniform moral precepts "taught" by television. The power of television as a means of social control has not escaped those who have appointed themselves as the elite moral guardians for the society at large. "Correct" environmental values are taught on everything from Saturday morning cartoons (e.g., *Captain Planet*) to the evening "news." De Tocqueville would marvel at how television has become the instrument for social control that he anticipated would be "minute, regular, provident, and mild."[14]

Blaming Genesis

De Tocqueville may have been less surprised than I about universities also becoming instruments for such social control. Just as we have observed in other nations, the discrediting of competing moral authorities is an important tool for those who see themselves as "teachers." Historian Lynn White, Jr., popularized growing disenchantment with the Judeo-Christian heritage when he charged it with being at the root of our modern environmental crisis.[15] A generation of young minds has uncritically absorbed his words and felt better about turning away from their earlier selves in search for "new selves" in non-traditional religions, mysticism, and re-enchantment of the earth with spirits of all kinds. Antipathy for traditional Jewish and Christian religion is now well established in academic circles, despite excellent Biblical scholarship refuting White's thesis and historical studies showing that China and other Eastern civilizations with more "nature-oriented" religions (such as Buddhism) have caused as much, or even more, disruption to forests and other natural environments.[16]

New "Community Of Nature"

Many of these environmental "teachers" reject the traditional idea of stewardship because it places humans above other forms of life, and impedes the development of a truly egalitarian vision in which all forms of life, including humans, are part of the same "community of nature"—a community in which no one species is any more important, or has any greater rights, than any other.[17] As a result, secular laws such as the Endangered Species Act take on special moral significance as instruments for enforcing "equal rights" in the new community of nature. "Private" selves accustomed to assuming stewardship responsibilities for the environment are transformed into "ecological" selves obliged to participate in nature as "just another species."

Trading Stewardship For Humility

By trading stewardship for humility, there is a deliberate rejection of the "old selves" that believed humans are morally and spiritually exceptional in the natural world. Hence, rejection of stewardship also implies a rejection of the traditional belief in the moral integrity of the human individual. Only those clinging to the "old ways" of individual privacy and "frontier mentality" would be so "arrogant" as to take responsibility for other forms of life, while those who follow the "new" way would seek "community with nature." Only a few—the "enlightened teachers"— have sufficient "humility" to understand nature and should be trusted with deciding how the land should be used. The rest should follow the "teachings" of the "philosopher-kings."

Moral Persuasion Brings A Lose-Lose Tragedy

I have written this book because continued reliance on moral exhortation cannot solve our environmental problems and will cost us our political freedoms. This is a lose-lose situation nobody wants. Americans are not hopeless victims of a society tyrannized by the politically correct. Americans are a strong people with proud traditions. The best way of resisting moral persuasion is to assertively take pride in who we are and to allay the fears upon which it feeds by taking full responsibility for the ecological problems we create. Affirmation of freedom and belief in ourselves will best prepare us to face unprecedented challenges in protecting and managing our environment.

Reject The "Fight With Ourselves"

But to succeed we must end the "civil war" that rages within us and between us. We must reclaim ownership of our selves, restore American political traditions, and take responsibility for solving the real environmental problems that confront us. This work can only begin by rejecting the struggle between the "old self" and the "new self." We must reject both Mao's dictum to "fight 'self'" and Gore's declaration of "war with ourselves." Protracted internal struggle can lead us down paths to tyranny by allowing a mass mentality to be substituted for individual dignity, caring, and responsibility. I know, because I am a seasoned veteran of this war.

Chapters 4 through 9 will describe how I rejected the declaration of "civil war" and reclaimed ownership of my self. I will chronicle how this war originated and how I have freed my self by taking responsibility for parts of myself I didn't want to accept. Permitting others to own my self was comfortable because it permitted me to hide from what I did not want to face. I have reaffirmed the roots of American political culture: firm beliefs and traditional institutions now protect me from the less desirable parts of myself. I have learned much, not the least of which is that Plato was dead wrong about human nature. And those who "teach" virtue and dream of utopias are just as wrong as Plato was. The road to hell is indeed paved with good intentions!

Moral Persuasion Is Fickle And Slippery

Moral persuasion is a *method* for getting people to voluntarily conform to the state or a powerful elite. It is not communist or capitalist, environmentally "green" or industrially "gray." Its independence from any particular ideology makes it hard to recognize, since we tend to tag things with ideological labels. Its growing use in the United States also shows that it does not necessarily require administrative coercion through "command and control." Because it is a flexible tool for exercising social control, moral persuasion is especially fickle and slippery. Consequently, we can most easily detect it by the way it stimulates or exploits a struggle between an "old self" and a "new self" and is hostile to individuals who refuse to give up ownership of themselves to the state, an elite, a corporation, a social movement, or a cult.

Moral Persuasion Has American Roots

Two fully American traits contribute to the success of moral persuasion: devotion to rationality and science and a Calvinist tendency to split people into the "pure" and the "impure," the "saved" and the "damned." An over-emphasis on the authority of science causes

Americans to look to "experts" for advice on issues that are fundamentally choices about right and wrong.

Scientists As Philosopher Kings

Some environmental scientists and ecologists have used their well-earned scientific reputations as justification for speaking as "moral authorities" who recommend which lands should be used and which preserved and who should use water, range lands, and forests and who should not. Their moral choices are presented to the public as "science-based plans." These plans are built on ethical and political choices for which scientists are unprepared. While it is productive for scientists to debate the ways in which science and moral issues get mixed up (as I am doing in this book), there is no way to hold scientists accountable when they occupy decision-making positions and present moral choices as scientific judgments.

I will refer to scientists who become decision-makers and mix moral choices with scientific judgments as "charismatic scientists" or "scientists." When I place quotation marks around "scientists," I am referring to people who exceed the authority of their well-earned scientific reputations by acting as *public decision-makers* and making moral judgments that have major consequences for people's lives. Scientific training and experience does not give people the necessary background, or authority, for making public policy decisions involving complex moral issues.

The *Seattle Times* recently carried an article entitled "Forestry In The Age Of The Philosopher King" featuring the ideas of Dr. Kai Lee.[18] Lee argues that democracy is not suitable for managing ecosystems, and that an ideal means for combining "excellence in science with success in politics, administration, and professional practice" would be "'philosopher-kings'...the wise and powerful governors of Plato's *Republic*."[19] He acknowledges that "...we should allow for [such] genius but not count upon it," by developing a partnership between "...the science of ecosystems and the political tasks of governing."[20]

Lee calls for idealism in science and naively assumes that scientists are capable of separating moral and scientific judgments when laying out alternative strategies for managing natural resources. He also assumes that moral decisions can be left to politicians who are skilled at working out compromises and can be held accountable for results. As is typical of many academics, Lee paints a very idealistic picture and ignores the realities of what scientists actually do when they apply their knowledge to practical problems. As will be shown in the next chapter, "scientists" involved in making recommendations for Pacific Northwest forests adopted the role of philosopher-kings, circumvented pluralistic democracy, placed themselves above the rule of law, and mixed moral and scientific judgments when defining problems, selecting evidence, and formulating management options. And they did not employ rigorous scientific hypothesis-testing procedures that would have protected the scientific community and the public from arbitrary imposition of these moral judgments and the distortions in "truth" they produced.

Uncritical Reliance On Calvinist Ethics

When mixing scientific and moral judgments, "scientists" may have uncritically embodied the Calvinist outlook so typical of Americans. I will also explore how the internal struggle between "good" and "evil" is in large part a product of Calvinism—a struggle that is resolved when what people do or fail to do is taken as a sign of their inner spiritual status. Hence, "they" are bad people because "they" "rape the forests," "destroy the wetlands," "wear furs," and "pollute the rivers."

Later chapters will discuss how we see in the faults of others what we refuse to see in ourselves. Yet the truth is that we all live in houses made of "murdered" trees on land "raped by ax and chain saw," eat food that has been "torn from the belly of Mother Earth," and drive or ride in vehicles that "choke the earth" in pollution. Accepting who we are as consuming organisms, respecting others, and striving for more responsible production and consumption is a choice we can all make to avoid a war with ourselves. We cannot learn to live sustainably with our surroundings until we grow beyond the divisive elements of Calvinism by taking responsibility for ourselves and our surroundings.

Restoring America's Political Strengths

Only by building on the strengths of our political traditions can we successfully address the serious environmental problems that we have created. Problems can be solved by taking responsibility for our behavior through limiting the control that centralized elites and government have over our lives, strengthening families and communities, protecting pluralistic associations, and reducing the role of government in science, professional practice, and religion. None of this can be achieved unless we reject the declaration of "war on ourselves."

We depend on people who cut forests, graze cattle, build roads and shopping centers, fish the oceans, or farm furs. These people are not our enemies, and, despite what is said by our revolutionary "teachers," they are not evil. They are simply doing what we have indirectly asked them to do when we consume products made from natural materials or occupy land from which forests have been cleared or water drained. These people are valued producers who satisfy our needs and sustain our way of life. We depend upon them, just as they depend upon us. This truth is the basis for learning how we can take greater responsibility for protecting and enhancing our environment.

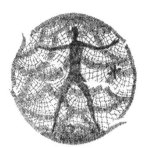

2

Obeying Ecological Authority

Moral people can be driven into committing immoral acts even though they know (or believe) the acts are immoral—providing that they are convinced that the experts (people who, by definition, know something they themselves do not know) have defined their actions as necessary.

Zygmunt Bauman
Modernity and the Holocaust

I had long known the answer to the question of why the suffering of rural producers was ignored. But I didn't have the courage to say it—even to myself. For years I preferred to believe that suffering could only be willingly inflicted by cruel people whose authoritarian character led them to persecute others. Like most of my contemporaries, I took comfort in my belief that I was not associated with such "bad people." Yet how wrong I was. Wrong not because I was surrounded by cruel people, but wrong because I denied the fact that under the right conditions most of us will inflict great suffering on others if we think it is necessary to do so. I denied the fact that most of us are well prepared to be cruel if authorities convince us it is necessary.

Few believed this possible until the 1970s when Stanley

Milgram, a Yale psychologist, terrified the social science research community by publishing results from experiments in which he demonstrated how easily subjects could be induced to make other people suffer.[21] The social science community has yet to fully absorb the implications of his results, since what he showed throws into question the long-accepted idea that morality is a product of society—that what is accepted as good and bad behavior is determined by the majority of people. Milgram demonstrated that social scientists can no longer comfortably assume that social authority is the legitimate basis for moral judgment, and that immoral actions are those that deviate from the standards promoted by a rational society. But first to his experiments.

Milgram was concerned with the way in which some people blindly obeyed authorities, and designed a series of experiments to evaluate the conditions under which people would obey. Volunteers were recruited to a psychological laboratory and asked to carry out a series of acts that increasingly conflicted with their consciences. The primary question was how far the volunteer would comply with the experimenter's instructions before refusing to carry out the required actions. It should be noted that rules protecting human subjects would make these experiments impossible in current research laboratories.

Two individuals, one designated a "teacher" and the other a "learner" were placed in a psychological laboratory (Milgram, apparently, used the word "teacher" without realizing its connotations in systems relying on moral persuasion). The experimenter explained that the study was evaluating the effects of punishment on learning. The learner was seated in a small room, with his arms strapped down to prevent excessive movement. An electrode was attached to the learner's wrist. The learner was informed that he was to learn a list of paired words; whenever he made an error he would be given electric shocks, and that the shocks would increase in intensity with each error.

The learner was an actor who did not actually receive any shocks. The "teacher" was the real experimental subject. The objective of the experiment was to see how far the teacher would go in obeying orders to inflict pain on a protesting subject. The teacher watched the learner being strapped into the chair and was then taken to the main experimental room and seated before an impressive shock generator.

The face of the generator had thirty switches ranging from 15 volts to 450 volts, and accompanying designators ranging from "SLIGHT SHOCK" to "DANGER—SEVERE SHOCK."

The teacher was instructed to administer the learning test to the subject strapped in the next room. When the learner responded correctly, the teacher was to move to the next pair of words. When the learner made a mistake, the teacher was to administer an electric shock, but to increase the intensity of the shocks with each incorrect answer.

The learner was trained to feign increasing discomfort to match the increasing intensity of shocks. The learner grunted at 75 volts, complained verbally at 120 volts, demanded release from the experiment at 150 volts, and, after growing increasingly vehement and emotional, issued an agonized scream at 285 volts. The suffering inflicted on the learner caused tremendous conflict in the teacher.

The suffering caused the teacher to want to quit the experiment, and almost all protested to the experimenter. The experimenter played the part of an authoritative expert on memory and learning, and ordered the teacher to continue whenever there was a hesitation to administer the next level of shock. The only way the teacher could get out of the situation was to defy the authority of the experimenter. The whole purpose of the experiment was to determine when, how, and under what conditions people would break with authority when faced with a clear moral imperative to continue with the experiment.

The results of the experiment were morally disturbing, and many other researchers initially refused to accept them. A majority of the people would go to great lengths to comply with expert authority, despite the stress induced by the conflict between what they knew was right and what they were asked to do. Many obeyed the experimenter by inflicting great pain, regardless of the victim's agony and pleas for release from the experiment.

Those who volunteered for the experiment were not sadists or cruel monsters recruited from the fringes of society. They were ordinary people drawn from the working, managerial, and professional classes. What made the results so disturbing was that these "teachers" were regular folks, just like the rest of us. Yet these "regular folks" had carried out actions incompatible with fundamental standards of morality by deferring to the authority of those who claimed to be scientific experts in learning. Milgram summarized the lesson of these results as follows: "...ordinary people, simply doing their jobs, and without any particular hostility on their part, can become agents in a terrible destructive process."

Obedience To Authority Is Widespread

Like so many people, I had believed blind obedience to authority was restricted to a cruel people (bad people), who, like Hitler's Nazis, could inflict great suffering if given a chance to do so. Milgram's experiments taught me that inhumanity can be a product of social relationships in which we often unknowingly become agents for cruelty, destruction, and evil. The Nazis were only the most extreme manifestation of a ubiquitous tendency to abandon one's self and blindly obey authorities whose motives are uncaring, cruel, or evil.

Ordinary soldiers have the capacity to carry out atrocities such as the My Lai massacre when ordered to do so. Incarceration of Japanese-Americans during World War II was a cruel act carried out by ordinary Americans following the orders of their government. Blind obedience is also evident in recent revelations that during the 1940s and 1950s the U.S. government experimented with its citizens by administering radioactive chemicals to unknowing subjects. But government is not the only authority ordinary people readily obey.

Cruelty is repeated almost daily in the workplace when superiors order their subordinates to force workers to resign by making their work life unpleasant. In families, parental or spousal authority is often used to prevent a sibling or spouse from reporting abusive treatment of other family members. All of us, in almost any daily social relationship, have the capacity to do cruel things to others if we are put in a hierarchical relationship where we look to authorities or experts for guidance on what to do. What is most troubling are those situations where we comply with authorities without realizing the suffering we have inflicted on others. Yet that is exactly the situation in which we find ourselves with regard to many environmental problems.

Respect For Scientific Authority

We turn for guidance to experts who are recognized authorities on ecological systems and environmental problems. The authority of science commands our respect, especially when we are unable to understand complex biological and physical processes and the technical jargon scientists use to describe them. Respect for scientific authority leads us to go along with the "scientist's" plans for conserving endangered species, preserving endangered ecosystems, protecting wetlands, and maintaining biological diversity. Yet implementation of these plans can cause extreme human suffering when people are put out of work and driven into poverty, publicly vilified for "destroying nature," lose the rights to earn an acceptable living from their property, are unable to use their land as planned, or are arrested and fined and/imprisoned for violating laws protecting nature by creating a minor disturbance in a wetland, cutting a tree, or building a road on an unstable slope.

One reason we so readily go along with these plans is because we have an amazing amount of trust in "scientists" and revere the authority of science as an institution. I will discuss some of the cultural and psychological dispositions to go along with the "scientist's" plans in the next chapter. But what I am most concerned with here are social relationships, such as trust in scientists or bureaucratic organization, that makes it easier for people to see human suffering as a necessary and unavoidable result of protecting the environment.

Making It Easier To Inflict Suffering

Milgram discovered that people find it easier to inflict suffering on others when the moral authority of expertise is asserted. When this happens greater social and physical distance is maintained, and a sequence of increasingly unpleasant actions traps people into doing things they know are wrong. Consequently, differences in opinion are suppressed. All of these conditions have at one time or another characterized reforms pushed by the environmental movement and implemented by government agencies.

Clinton's "Forest Conference"

President Clinton's plan to conserve forests in the Pacific Northwest was developed through a process that embodied almost all of these conditions.[22] He appointed a multi-agency Forest Ecosystem Management Assessment Team (FEMAT), and charged it with developing a plan for resolving the political gridlock over the management of federal forests and surrounding lands.

President Clinton would fulfill his campaign promise to hold a "timber summit" (renamed a "Forest Conference") by meeting publicly with key local and regional leaders. The "Forest Conference" would involve the President, Vice President and five Cabinet members and would be held in Portland, Oregon on April 2, 1993. Fifty participants who had been involved in political, technical, or scientific aspects of the conflict were to be invited to present ideas and join in the discussion.

I was honored to be one of two sociologists to invited to sit at the table with the President, Vice President, and key departmental secretaries. Along with many other participants, I left Portland filled with pride in my accomplishments and confidence that national leaders were finally listening to all sides of the debate. I felt the President was sincere, and I still want to believe in his sincerity.

FEMAT: The Real Agenda

But the President was not well-served by his staff. The Forest Conference soon took on its real meaning as an orchestrated public relations event to make people feel they mattered and convince the public that the President was fair-minded. The real agenda was FEMAT, and it had been set up weeks in advance and was designed to involve only those who favored the Clinton Administration's version of "ecosystem management." FEMAT was led by "scientists," most of whom had backgrounds in forestry.

Ignoring Human Suffering

Several weeks after the Forest Conference, I was invited to advise the FEMAT social assessment sub-team. However, I soon resigned when I learned that FEMAT did not want and would not use any information that would not support the long-term goal of restoring forest to their pre-settlement conditions (how forests appeared when the first European's arrived). FEMAT "scientists" did not request and would not accept the substantial amount of information I had assembled on the human suffering already caused by sudden federal timber harvest reductions. Moreover, they were not interested in my estimates of the severity and extensiveness of future social impacts likely to result from further harvest reductions, and had no interest in social risk assessment methodologies that could help assess alternative management options or show how both species conservation and timber harvesting could be achieved. [23]. I wrote letters to President Clinton and the FEMAT social assessment sub-team to protest the non-scientific and unprofessional practices of excluding unwanted advisors and information.

Lack Of Professional Authority
The breach of professional ethics exhibited by the FEMAT process was avoided by almost all of my professional associates. The Society of American Foresters turned its head the other way, in part because two of its close associates and former policy advisors participated in designing the Forest Conference and the FEMAT process. My associates in sociological circles refused to acknowledge that there was anything wrong with a "scientific" committee that selected facts to support its preferences. The only voices calling "foul" were forest products industry associations and leaders of grass-roots community groups. Since no professional societies would stand up to the abuse of the scientific method or breach of professional ethics, I reluctantly joined with a colleague who had been systematically excluded from the FEMAT process and provided affidavits for the forest product industry associations that challenged the legality of FEMAT's exclusionary practices.

Charismatic Ecologists Are Above The Law
The exclusionary behavior of FEMAT "scientists" was challenged in federal court.[24] On March 21, 1994, U.S. District Judge Thomas Penfield Jackson ruled that the scientists preparing the FEMAT report had violated the Federal Advisory Committee Act by failing to open its proceedings to the public and by selecting information from a

limited circle of scientists. Jackson did not enjoin the President's draft plan, since he lacked the authority to tell the President how or where he could seek advice. However, Jackson did invite the plaintiffs to challenge the plan under a separate statute if it was adopted by the Administration. Administration officials indicate they will ignore the illegality of FEMAT and go ahead with the forest plan developed from FEMAT. Industry associations will undoubtedly challenge the legality of the plan.

The way FEMAT organized itself and conducted its activities illustrated many of the social relationships Milgram found were associated with blind obedience to authority: a team of federal and university "scientists" retreated behind closed doors, maintained complete secrecy, excluded both unwelcome ideas and scientists with whom they disagreed, came to see themselves as an authoritative and self-directed group, and developed a plan endorsed by the President.

Most importantly, the tremendous human suffering that would be caused by the plan was ignored because everyone in FEMAT deferred to ecological authority and was convinced that it was necessary to drastically reduce timber harvesting. Now that ecological authority has prevailed, the reality of human suffering is becoming evident and is an intolerable burden for citizens and local officials. A recent newspaper article[25] reported that according to a school superintendent in Morton, Washington, "More than 33,000 children in Lewis, Mason, Grays Harbor, and Pacific counties urgently need help." Yet this is only a small fraction of the three-state region impacted by the President's plan, and it doesn't even mention the financial and emotional stress of the parents whose lives have been torn apart by federal timber supply decisions.

The actions of officials in the Clinton Administration is consistent with the belief that "charismatic ecological scientists" should be given the authority of philosopher-kings. The Administration has given these "scientists" the authority to disregard democratic pluralism, including laws guaranteeing an open society. Such action flaunts the moral and constitutional principles upon which the United States was founded by placing "scientists" above the law. The politics of science in the FEMAT process resembled the conflict between "Reds" and "experts" in the Chinese bureaucracy, only in this case the conflict is between the "Greens" ("scientists") and the "experts" (independent scientists who participated in or were excluded from FEMAT).

Science Is Claiming Moral Authority

FEMAT's use of science was not new to government-controlled research. The U.S. Forest Service, in particular, has a long history of selecting information that will support its policy preferences and discrediting or ignoring information that threatens these preferences. It spent over 50 years trying to prove that fire was an alien, destructive threat to forests, only to reluctantly admit that it had been wrong.[26] The universal application of clearcutting to all forest types in the United States (even though it was suited to some regions) was supported by similar commitments to policy preferences. FEMAT's selection of facts to justify major reductions of timber harvesting suggest that we are looking at the same horse with new riders. All that has changed is the exceptional zeal of the cause. "Saving the earth" is a stronger justification for politicizing science than saving the forests from fire.

Politicization of scientific authority, not political will, has been increasingly relied upon to direct the administration of environmental laws. This is most clearly expressed in the dominant role that "scientists" have played in developing plans to protect the old-growth forests of the Pacific Northwest. But when unchecked by independent moral or professional authorities, exclusive reliance on politically-motivated scientific authority traps administrators in unpleasant actions that ultimately lead them to try to limit open discussion and dissent.

Trusting The Expert

Milgram drew on the authority of science to convince subjects that, no matter how troubling, their actions were required. The moral authority of the expert was used to reassure subjects when they expressed moral anguish over their actions. Most subjects deferred to the ethical judgments of "higher authority" who routinely reassured them that "No permanent damage to the tissue will be caused." They trusted "the expert" to tell them what was ethical and unethical, right and wrong.

A few "scientists" have assumed a similar role in contemporary environmental conflicts, and most citizens, like Milgram's subjects, fail to understand that "scientists" are poorly prepared to advise people on ethical choices. Science epitomizes the rational separation of means from ends, and specializes in understanding how and why things happen. Unlike questions of means, ends are moral questions that involve justifications for what should or should not be permitted.

Scientists Mix Ethics And Facts

Moral justifications are questions for ethicists, theologians, and citizens participating in democratic processes, not questions for scientists. Yet the fragmentation of modern, bureaucratic organization allows "scientists" to mix ethical and factual judgments—even to the point of imposing their own preferences on the majority by using the authority of science to justify an ethical choice. "Scientists" have been encouraged to use their scientific authority as justification for exercising moral leadership because Congress and administrative agencies (such as the U.S. Forest Service which administers most of the federal forests in the Pacific Northwest) have so fragmented action that they have difficulty pursuing a purpose.

Out Of Touch With Real People

Yet these same qualities of organizations protect their members from the realities of moral choices. They insulate experts from the consequences of their actions and allow citizens to be treated as objects to be manipulated rather than as clients to be served. As modern warfare has shown, it is easy to inflict suffering on someone we can neither see nor hear. It is harder to harm someone we can hear but not see, and still harder to harm someone we can see at a distance. It is most difficult to harm someone we can reach out and touch.

Milgram found that only 30 percent of the subjects in his experiment continued to follow the commands until the end of the experiment when they were told to force the victim's hand to the plate through which they received the electric shock. Obedience increased to 40 percent when subjects were asked to manipulate levers on the control desk in the presence of the victim. When the victims were hidden behind a wall, and subjects could only hear their anguished screams through a sound system, obedience increased to 62.5 percent. Turning off the sound system increased obedience only to 65 percent—suggesting that seeing and touching are the primary ways we maintain human feelings for others. Milgram concluded by remarking that "In modern society others stand between us and the final destructive act to which we contribute."

Bureaucracies Put People At A Distance

Creation of physical distances between people and the consequences of their actions is one of the most noteworthy achievements of modern society. Modern societies are masters of bureaucratic organizations. Bureaucracy splits actions into separate stages and sets each stage apart from the next through the hierarchical arrangement of authority and specialization of tasks that are assigned to each functionary in the hierarchy. The process of rationalizing actions makes it easier to contribute to human suffering. The effects of physical distance is even more pronounced when people who occupy specialized offices are spread across a region or nation and are responsible for tasks affecting people or things at great distances from them.

Bureaucratic organization characterizes most state and federal agencies authorized to address environmental problems. In addition, each agency is given authority for specialized issues or problems. The U.S. Fish and Wildlife Service has the authority to administer the Endangered Species Act that was designed to protect individual species of plants and animals. Both the U.S. Forest Service and Bureau of Land Management administer federal lands on which live many threatened or endangered species. State agencies have been given the authority to police the use of private lands on which many threatened or endangered species are found. Other federal and state agencies are assigned authorities and responsibilities for regulating wetlands, air quality, water quality, and toxic wastes. Health and welfare agencies are responsible for meeting the basic needs of people displaced by species protection plans.

All this creates a complex web of detached functionaries who have very little, if any, idea how their separate actions actually affect people—or even affect the natural environment they are charged to protect. The perverse genius of rational organization is that it makes it possible for people to make life-threatening decisions and yet remain at peace with themselves. What is most amazing is that they often do not know what they are doing, yet are not uncomfortable with this situation. Such moral blindness is a necessity for fighting a war, but it cuts people off from the human suffering they create when their objective is to integrate human activity with ecological processes, not to defeat an enemy.

Turning People Into Objects

The physical distance Milgram observed between the subject and the lonely victim sitting behind the wall was coupled with a sense of social distance created by treating the victim as an *object* to be manipulated. A group feeling was contrived by maintaining physical closeness and a sense of cooperation between the subject and the experimenter. A feeling of an in-group was created in the short period of the experiment, and the victim was excluded from any recognition as a participant with a voice in the experiment. The learner was an *outsider* whose function was to help advance understanding of memory and learning by doing as he was told and suffering the shocks.

Names that define people as different or outsiders make it easier for them to be treated as objects to be manipulated. The experimental "subject" is not a real person. Using words to turn people into objects is common in wartime when the enemy is referred to as "Gooks," "Japs," "Gringos," or "Krauts." But names are also common in racial, gender, and lifestyle stereotyping, and in social movements that pit people against one another.

The structure of bureaucracies make them adept at turning people into objects, since experts who are assigned discrete tasks to treat parts of an overall problem only see abstractions of real people—they tend to see stereotypes, not full personalities with aspirations and problems like the rest of us. Hence, as I will explain in subsequent chapters, loggers are experienced only as stereotypes—"buffalo hunters," or "land rapers,"—not as complex men with wives, children, debts, dreams, sins, fears, hopes, and pride in their accomplishments.Similarly, environmentalists are often stereotyped as "enviros."

Humans Are "Just Another Species"

Adoption of a detached, scientific perspective makes it easy for many scientists to turn people into objects. For some "scientists," humans may be treated as "just another species" to be subjected to the laws of nature; widespread human suffering may thus be seen as a "natural, hence justifiable, consequence" of excessive population growth.[27] Detached perspectives on environmental problems are never neutral scientific positions, since they embody moral judgments about the relative importance of people and other forms of life. These questions of morality and science will be more fully explored in the next chapter. But before turning to them, I will explore the way a monopoly of scientific authority traps people in an unpleasant course of action.

Getting Sucked In

Milgram also observed that subjects found it difficult to withdraw from the experiment because they got trapped in a sequence of actions in which each level of shock was only a small step beyond the previous level. Since there was no communication with authorities other than the scientific expert, subjects found it difficult to tell when they had gone too far. They lacked a clear line dividing permissible and impermissible action. Moreover, the deeper they got into the experiment, the more they felt that a decision to stop would imply that their previous step might also have been wrong to administer. The longer they participated in the experiment, the more they became committed to the moral position articulated by the experimenter. They were progressively trapped in a net they helped make by succumbing to the exclusive authority of the expert. The uncompromising authority of the expert set up an internal tension between their conscience and their inclination to obey. They quickly slipped into the role of "teacher."

Scientists may come to monopolize moral authority by getting trapped in a sequence of moral judgments from which they cannot escape without risking their reputations as experts. The principal "scientists" involved in developing President Clinton's forest plan for the Pacific Northwest fully assumed they had the authority to make moral choices that would inflict enormous suffering on individuals, families, and communities. FEMAT was the fourth in a sequence of "scientific" plans that all involved significant moral judgments about the welfare of people and the survival of biological populations.

This sequence began in 1989 with a committee that developed a conservation strategy for the northern spotted owl.[28] The committee report led to the listing of the northern spotted owl as an threatened species. Shortly after issuance of the owl conservation report, another team of forest "scientists" was asked to develop a report on the management of old-growth forests in the Pacific Northwest.[29] They, unashamedly, nicknamed this document the "Gang of Four Report." Still another team of "scientists" was asked to study the viability of species associated with old-growth forests in the Pacific Northwest.[30] The report from this study was issued about the time President Clinton requested establishment of FEMAT.

As soon as FEMAT was completed, many of the same "scientists" immediately engaged in replications of the FEMAT process in the Sierra Nevada forests of California and the East Side forests of Washington and Oregon. Further extensions of the FEMAT process to other regions of the country are planned.

Trapped In The Myth Of "Science-Based" Plans

"Scientists" were selected for these assignments because of their outstanding reputations as specialists. They appear to have trapped themselves into making moral judgments by rationalizing each moral judgment as only a small step beyond the previous judgment. An admission that they might have been wrong in making a moral judgment for which they were inappropriately trained and totally unprepared became more difficult with each step because they would have to admit that all previous judgments had similarly exceeded their specialized expertise. Like Milgram's "teachers," they were progressively trapped in a process from which they could not escape. They appealed to their authority as scientists to justify their moral judgments. This monopoly on moral authority convinced many in the public and the press to believe these were "science-based" plans and to ignore both the implicit moral judgments they embodied and the highly conjectural scientific judgments they contained.

Eliminating Disagreement

Milgram was concerned about the effects of such monolithic authority, as most of his experiments had been contrived to allow only the experimenter to communicate with the subject. He added a number of experiments in which more than one experimenter was present and the experimenters were instructed to argue about the command. The results were dramatic. Blind obedience could not be elicited, and subjects refused to engage in actions they did not personally agree with. Immoral actions were paralyzed by pluralistic debate. Pluralistic debate allowed the subjects to maintain moral ownership of their selves.

The effects of disagreement among experts were so convincing that Milgram did not pursue other possible combinations of authorities in his experiment. While the results might have been the same if he had introduced debate with an ethicist, a theologian, or a review panel constituted by ordinary people, a debate between scientific authority and authorities on morality might have had very different effects on how subjects would respond to a monopoly of scientific authority in a subsequent experiment. This would have made his experiment far more realistic by allowing the subject to learn from experience and even possibly develop both a strong conscience and the capacity for independently assessing a situation and making moral judgments about whether they wanted to participate.

Creating A Monopoly On Moral Authority

Moral authority has been monopolized by a few "scientists" in most debates over environmental issues. Participants in these debates tend to be "scientists," the courts, administrative agencies, and contesting interests (including industries, landowners, public land users, conservationists, and advocates for preservation). The courts have relied on scientific reports to tell them how to interpret substantive issues such as the maintenance of biological diversity. Powerful environmental interest groups are so uncompromising that Congress has been unwilling to exercise its authority to clarify or reconstitute the law. Religious institutions have tended to remain on the periphery. But when they have become involved, they have either found company with the moral persuasion of apocalyptic environmentalists or ministered to those who suffer poverty and hunger resulting from environmental preservation. Most ethicists or theologians who have become involved publicly are those who advocate the environmental preservation agenda. The national and urban press tends, with a few exceptions, to bow at the altar of charismatic "scientific" authorities and add their own versions of moral exhortation.

As a result, there has been little meaningful debate between competing scientific and moral authorities. Like Milgram's subjects, both government functionaries and citizens are trapped in a sequence of increasingly unpleasant decisions from which there seems to be no escape. A few ecological "scientists" are succeeding in trapping people in a monopoly on "ecological morality." Massive social and economic disruption seems inevitable if scientifically justified plans are to be implemented. Human suffering is seen as regrettable, but necessary in order to protect nature.

Other possibilities for protecting the environment without inflicting so much suffering are not considered because the majority has willingly acknowledged the "scientist's" right to decide what form of environmental protection is "best." The "scientist's" role in making arbitrary moral choices has been clouded by the almost total authority they have been given. The public has been hoodwinked into thinking that the "scientists" had simply described what had already taken place, and that rational people should wake up and adjust to it. Nothing could be further from the truth, and nothing could be more abusive of a truly scientific approach to these problems. Science demands chronic doubt, not obedience to authority.

Silencing Pluralistic Debate

Hence, advocates for a free society with open exchange of views were not surprised when the President Clinton's plan for Pacific Northwest forests was not well received by any of the interests whose gridlocked conflict the plan was supposed to resolve. Also not surprising, given Milgram's findings, was the suggestion by some FEMAT "scientists" that gridlock could not be resolved until the pluralist political system was short-circuited by creating a "cooperative partnership" between federal "scientists" and local stakeholder groups (creating

exclusive relationships between experimenters and local subjects)—
the basis for the experimental "Adaptive Management Areas" described
in Option 9 of the President's plan.[30] From the viewpoint of philosopher-kings, their assessment of pluralism was entirely "correct."
Democracy is cumbersome and requires openness and compromise.
The monopolistic authority of science cannot be implemented until pluralistic debate has been silenced. Well-meaning "scientists" opened the
doors to top-down control without fully recognizing what had they had
done.

Experts As Functionaries For Power

Milgram said that subjects become *agents* when they come to
see themselves as responsible for carrying out the wishes of a superior
authority. Obedient subjects were good agents for his experimenter.
David Bella, an engineer at Oregon State University, talks about this
problem in terms of "functionaries": a functionary is "one who defines
responsibility in terms of one's own assignment and avoids the responsibility for questioning the validity of such assignments or the ends they
serve."[32]

Like many in the scientific community, Bella argued for scientists to maintain *independent* disciplinary communities. Independent
communities of scientists are essential, Bella said, for eliciting public
confidence by rejecting "false information, inappropriate assumptions,
and mistaken perceptions."[33] Public trust in scientists depends on how
well scientists discipline themselves. Bella cautioned that

> Without disciplinary communities that sustain independent
> inquiries, criticisms, and debates, the evolution of professional
> knowledge falls under the selective influence of society's dom-
> inant organizational complexes. If this occurs, then engineers
> (and scientists) are likely to be seen as mere functionaries in
> the service of power.[34]

Strengths Of Moral Autonomy

Milgram was aware of the way such functionaries blindly obey
authority, and contrasted *autonomy* with agency. He said that
autonomous persons see themselves as independent moral agents who

answer to themselves. Autonomy implies that individuals develop the capacity for making independent moral and scientific judgments.

This is where Milgram's findings are most disturbing to other social scientists. Sociologists have long assumed that immoral behavior is the result of basic animal passions—pre-social or anti-social drives—and that moral behavior results from social conditioning and teaching of virtue. The forces of a rational society are seen as necessary to lift people out of barbarity. Morality is seen as a product of *society*, not the individual. As a result, people who resist the standards promoted by society inevitably fall into immoral behavior. Most of our practices for educating individuals and correcting deviant behavior are based on this assumption.

Society Can Make Good People Do Evil Things

Milgram upset the social science apple cart when he showed that immoral behavior can be the product of society. He demonstrated that good people could be trapped into doing evil things when placed in authoritarian social relationships. Even more disturbing, he showed that the independent moral self is something some societies try to silence, exploit, re-direct, or remove from individuals. He showed that society was as capable of becoming an agent for evil as it was of promoting good. The most important conclusion to be drawn from his work is that morality is fundamentally an *individual responsibility*, and only secondarily the responsibility of society at large. A good society is the product of morally autonomous citizens who adhere to sound principles that discipline their behavior. And, as Tocqueville observed, morality is acquired in autonomous families, small communities, local religious gatherings and groups of believers, and voluntary associations.

Responsibility For Other People And Nature

Some sociologists have taken moral autonomy as the cornerstone for a new theory of morality. Zygmunt Bauman[35] has defined "responsibility for the other" as the foundation for morality, and located responsibility in the moral sensibilities of autonomous individuals. As will be developed in the next chapter, I will put more emphasis on the need for highly localized social groups to teach individuals to be accountable for respecting the dignity of all people. I will also add "responsibility for nature" to this definition of morality.

Hence, I will close this book by calling for restoration of political institutions that 1) reduce social and physical distances between experts and citizens, 2) create realistic expectations about what scientists know and do not (or cannot) know, and empower citizens with what they need to ask of scientists to assure that scientists are held accountable to the scientific method, 3) call upon ethicists, theologians, citizen groups, and political representatives to break the monopoly on moral authority recently claimed by ecological "scientists," and 4) cultivate active exchange of viewpoints among a wide variety of interested participants and stakeholders.

But responsible living also requires that we govern ourselves by institutions that demand accountability of individuals as to how they affect other people and the natural world upon which we all depend. It will do no good to free ourselves from social arrangements demanding our blind obedience unless we are guided by principles giving us a more inclusive sense of moral responsibility for the environment.

3

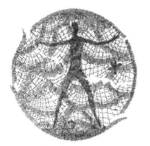

Beyond
Calvin's
Ghost

What Calvinists proclaim in the name of God, Social Darwinists assert in the name of a strangely providential Nature.

Richard L. Rubenstein
The Age of Triage

I was once fond of telling close friends who advocated radical environmental reforms that they were "closet Creationists." I decoded my cryptic remark by explaining, "Although you disavow all traditional religious beliefs in the Creation, you think the same as those who believe in the Creation."

As they grew increasingly uncomfortable with me, I quickly got to the punch line, "You talk about people as if they were not part of nature, as if they are alien creatures that defile nature by their very presence. Moreover, you have your own version of original sin, and, along with many fundamentalist preachers, rail on and on about people being a 'cancer on the Earth.' If you believe in evolution, as you profess to, then you are contradicting yourself by talking about people as separate from nature. Whatever people do must be natural, and hence 'good,' if we evolved along with the rest of life."

I was never successful at getting people to talk about what I saw as a contradiction. Then one day it dawned on me why I could never get people to respond. They were actually as religiously Calvinistic as many whom they opposed, they just had a different version of who was damned and who was saved—who were the elect and who were the unwashed masses. And they had no problem in thinking of people as alien to nature because the masses were impure and did not belong to the elect few who shared special knowledge, humility and love for nature. As the elect, they shared a moral community with nature to which others did not belong and could not belong.

What had first appeared as a contradiction now made perfect sense. Other people (certainly not themselves) were a "cancer devouring the earth." That is why they could drive cars without feeling guilty for polluting the air or condemn loggers for cutting old growth trees and at the same time buy redwood, cedar, or Douglas fir lumber cut from old-growth trees to remodel or repair their homes.

Calvinism Is A Backbone Of American Culture

Calvinist thinking has been a backbone of American culture since colonial times.[36] It has been the basic theory of life in one social movement after another. Most central to Calvinist thinking is the theory of predestination. Some people, the elect, were predestined for eternal salvation, while the remainder were predestined for eternal damnation. Few theories could have made more sense to those engaged in the harsh realities of developing a frontier nation. And few theories could make more sense today as people struggle with the problems of meeting the needs of exploding populations while also protecting nature.

But what made Calvinism most influential is the tendency to view worldly success as a sign of divine election and worldly failure as a sign of eternal damnation. The poor, the disadvantaged, and the marginal members of society are not only social outcasts but are condemned by God as well. Therefore, it is thought they deserve the suffering they must endure. Those who are either unable to help themselves or to respond effectively to the help of others are destined to suffer, so their suffering should not be seen as a serious social problem. It is ordained, or "natural."

Calvinism Was Secularized

Not long after colonial settlement, Calvinism spread from Protestant churches to the secular society, and is now part of common American culture. Adherence to the Protestant work ethic still serves as a sign for certifying the elect. Much debate over welfare policies and the treatment of the homeless revolve around the issue of whether people who rely on the government or charitable giving are willing or able to work. A willingness and ability to work places one inside the dominant national community.

Social Darwinism Mirrors Calvinism

Yet there is another side to Calvinism that often goes unnoticed. This is its connection with Darwinism, especially social Darwinism.[37] Most people don't see the connection of Darwinism and Calvinism because they correctly see that Charles Darwin's theory of evolution is in conflict with the Scripture's account of how humans originated. But Darwin's arguments about the "survival of the fittest" are a close fit with Calvin's theory that "only the elect will be saved." It was not accidental that Darwin's theories were more widely applied to social and economic life in the United States than anywhere else. For it was here that the "struggle for existence" was given moral significance and played out in the development of the nation's economy.

The full title of Darwin's classic work was *On the Origin of the Species by Means of Natural Selection, or the Preservation of Favored Races in the Struggle for Life.* The "favored races" of Darwin and the "elect" of Calvin were often thought of as the same people in late 19th and early 20th century America. Major industrialists, from the post-Civil War boom through the recent negotiation of the North American Free Trade Agreement, have talked about the "survival of the fittest" when they justified absorption of smaller businesses by larger ones and advocated competition in the world market economy. Survival in a job or in business was a sign of moral certification—that one was a "good" worker or entrepreneur.

"Nature" Is The Social Darwinist's God

Hence, I was not surprised to hear committed environmentalists, many of whom have embraced Darwin's biological theories in exchange for Christian doctrines, talk about people in much the same terms as American industrialists. Both talked about "competition" and "struggles for existence." But they have very different versions of who is the "elect" and who is the "damned"—and very different ideas about whether God or a providential "Nature" makes these decisions. What God, or the economic "profit" is to Calvinists, "Nature" is to social Darwinists.[38] God, or opportunity for profit-making, is the originating power for Calvinists, while Nature is the originating power for social Darwinists.

Everything Is Either Pure Or Impure

Calvinism and social Darwinism are so closely related, and so prevalent in the way Americans think, that other ways of thinking are obscured. This is best illustrated by the tendency for people to reduce conflicts over the environment to simple choices such as "economic growth versus the environment" or "jobs versus owls." Both sides of the argument seem satisfied with this division. The Calvinism (or Social Darwinism) of the conservatives takes one side of such issues, while the Social Darwinism (or Calvinism) of radical environmentalism takes the other side. Ronald Reagan, George Bush, Pat Robertson, and Rush Limbaugh are simply the mirror images of Al Gore, Bruce Babbitt, Jay Hair, and those who minister to the Earth or worship the Greek earth goddess, Gaia. Such split thinking does not permit a middle ground in which jobs, economic growth, and environmental protection are all served.

Calvin lives on in American culture. His ghost continues to play strange games with our minds by leading us to divide fellow humans into the elect and the damned. Social Darwinism continues to serve as a scientific justification for such splitting, and in an increasingly secularized society, becomes a legitimate substitute for Calvinism. Both ideas, either separately, but especially in combination, lead us to deny the human dignity and moral integrity of other people. Social conflict is an inevitable result of denying the humanity of others. But it also contributes to the "struggle" in our selves.

Radical Environmentalists Resemble Reaganites

Much of environmental conflict can be traced to the mental splitting associated with Calvinism and/or Social Darwinism. When people suffer as a result of the implementation of environmental protections, it is usually because their rights were ignored and they were excluded from normal considerations of social justice that would bring sympathy, compensation for losses, and remedial help. This is either Calvinist or Social Darwinist ethics at work. The victims are generally portrayed as "bad people who deserve to suffer for hurting the environment" (e.g., loggers or developers), or their suffering is justified as inevitable and necessary in the on-going "struggle for survival" (an increasingly frequent rationalization popularized by the Reagan Presidency).

Leaving Calvinism And Social Darwinism Behind

I don't think we can begin to resolve environmental conflicts and develop more sustainable relationships with our surroundings until we free ourselves from the simplistic thinking we inherited from Calvinism and social Darwinism.

Paths to living more responsibly with our environment will be charted by leaving Calvin's exclusionary imagery behind and finding new principles to guide us. Responsibility for other people and our surroundings offers us a place to begin in seeking more inclusive moral principles. Social Darwinism will be returned to the dusty book shelves of history when more humane principles are embraced.

It would be terribly presumptuous for me to suggest moral principles. I am neither an ethicist, theologian, philosopher, nor statesman. As a scholar, I can help simplify, organize, and present what others have said. But perhaps most importantly, I can share my experiences and rule out dead ends where there is little useful to be found and often hazards to be avoided.

Responsibility For What?

I have spend over 25 years studying human relationships to natural surroundings. Much that is important about what I have learned can be summarized in a simple diagram. Responsibility for other people is portrayed across the diagram, while responsibility for the environment is portrayed down the diagram. By responsibility for others, I mean that individuals learn to *hold themselves accountable for helping to meet the needs or welfare of other people*. Responsibility implies reciprocity—give and take—or a willingness to follow rules that may inflict on oneself what one is about to inflict on others.

By responsibility for the environment, I mean that individuals learn to *hold themselves accountable for how they affect their surroundings*, especially the ecological system upon which they, and others, depend. They follow rules requiring them to consider what changes in nature they are likely to make and whether these changes will cause irreversible damage or harm, or will have indirect effects on others.

People are either accountable or not accountable for other people or the environment. This results in four possible combinations of accountability. People who do not hold themselves accountable for what happens to others or the environment (Cell 1) can be considered self-centered, or "egocentric." People who hold themselves accountable for what happens to others, but not the environment (Cell 2) are considered human-centered, or "homocentric."

		Accountable for Other People	
		No	Yes
Accountable for Nature	No	Egocentric	Homocentric
	Yes	Biocentric	Ontocentric

By contrast, people who consider themselves accountable for happens to the environment, but not other people (Cell 3) can be referred to as biologically centered, or "biocentric." A biocentric ethic is the dominant outlook of most environmental preservationists. Finally, people who see themselves accountable to both other people and the environment (Cell 4) can be referred to as centered on all beings, or, to use the root meaning of *being,* "ontocentric."

Extending Responsibility To Other People
Our society has progressed from the more egocentric thinking of Calvinism and Social Darwinism, and struggled to find ways of exercising greater responsibility for other people. The elimination of slavery, extension of suffrage to women, and guarantees of equal opportunity to all people illustrate this progress. The widespread support for these reforms shows the willingness of people to consider the needs of others along with their own needs. Both liberals and conservatives have endorsed this extension of responsibility.

Extending Responsibility To Our Surroundings
But only recently have we begun to seriously consider how to become more responsible for our surroundings. The environmental movement has spearheaded the drive to take more responsibility for the environment. But in doing so it has tended to emphasize the needs of other species or the integrity of ecological systems and to de-emphasize the needs of people. Radical environmentalists have looked upon people as a threat to the environment. Emphasis on our surroundings and the exclusion of people has resulted in the development of a strong biocentric ethic—a view that the protection of other species or natural ecological systems from people is ultimately a good thing. Moral persuasion is serving increasingly centralized biocentric goals.

Biocentric Cult Of The Elect
Biocentrism has found friendly company with Calvin's ghost. What some see as the "elitism" of the environmental movement could be more accurately described in terms of a cult of the preservationist "elect." These are contemporary Calvinists, even though they may dis-

avow any affiliation with established religion. They believe their humility, sensitivity, and concern for the environment sets them apart from masses of environmentally unconcerned. But, unlike classical Calvinists, they have an ambivalent attitude toward sin. They believe some sinners have the potential for moral transformation and can be "saved" by declaring their commitment to " Save the Earth."

But, most importantly, their god is "Nature" or "the Earth," not the God of Abraham. Just as Social Darwinists located power in natural laws of competition, or secular economic Calvinists came to see power in the "free market," so biocentrists locate power in "Nature," "Mother Earth," "Gaia" (the Greek earth goddess), or "ecosystems."

Radical Environmentalists Are American

There is a tendency for Americans to be easily seduced by biocentric views because we still hold a Calvinist outlook on life, and have difficulty accepting responsibility for others. We tend to see ourselves through a lens of religious imagery as a favored people whose destiny is to lead the world to freedom and enlightenment.[39] Hence, it would be totally wrong to say that radical environmentalists are "anti-American." They are instead about as American as they can be. Environmental preservationists are simply the latest in a long line of reformers who have drawn upon our Puritan origins and Calvinist ethics to divide the world into the saved and the damned. The religious imagery we inherited from Calvin, although now highly secularized, is what makes the environmental movement so appealing. Calvinism is a familiar resting place in a time of uncertainty and rapid change.

And yet, this same religious imagery can lead to futile and destructive social conflict if we allow ourselves to be directed by the division of humanity into the saved and the damned. The biocentric viewpoint tells us there are too many people for the earth to support and that there are people whose occupations cause them to "destroy the earth." Calvinism tells us some people are morally corrupt and others are bearers of virtue and justice.

Haunted By Calvin's Ghost

Although seldom seen or felt, Calvin's ghost haunts attempts to promote environmental reform in the United States. The outline of Calvin's ghost can be seen in attempts by powerful environmentalists to convince others *they* should decide who *deserves* to live well and who does not. Calvinism is hidden in the idea that the powerful environmental leaders should plan *our* future by giving biocentric "scientists" the authority to decide how we will live with nature—how we will "manage ecosystems."

I have written this book because too few people are talking about the hidden forces that shape the environmental preservationist agenda and its implementation. I didn't see these hidden forces until Richard Rubenstein taught me how to identify Calvin's ghost.[40] Now that I have seen Calvin's ghost, I am especially concerned when I hear talk about "surplus human population," "land rapers," and "destroyers of ecosystems." Environmental preservationism may not know it is haunted by cultural themes that could, when taken to an extreme, turn it into a monster which would decide who would live and who would not. Anyone who has seen Calvin's ghost will conclude that biocentrism embodies moral exclusion and could, under adverse political circumstances, lead to horrors few would have anticipated. A secularized Calvinism provides no limits to radical solutions for environmental problems.

Most of us find it difficult to accept that Calvin's ghost may haunt a wide spectrum of American boys or girls who have uncritically adopted extreme biocentric ideas. We deny our fears by placing faith in the inherent "goodness" of the American people. Yet, while few all-American boys or girls would have the stomach to inflict violence, they certainly can help prepare the way for others to use violence by innocently undermining traditions that protect the moral integrity of the individual. And like Stanley Milgram's subjects, most American boys and girls would voluntarily follow orders to " Save the Earth."

Youth Is Naive And Unprepared For Life

I see this tendency in the forestry, conservation, and wildlife undergraduates in my classes (thoroughly American young people—

although not as ethnically diverse as the population at large). Over the last three years I have asked students to critique a radical environmentalist comic book developed for use in elementary school classrooms. The comic is entitled *Stop the Chop!,* and tells a story of heroic "Earth Rangers" who go the Oregon to stop "Zack Ax" and his fellow loggers from cutting old-growth forests.[41] When the loggers (stereotyped as fat, dumb, ugly, and greedy) protest their arrest by producing a government contract giving permission to cut the trees, the Earth Rangers respond "We represent something bigger than the government, Zack Ax." "Yeah! Earth!"

Less than one in ten students see the police state tactics of the Earth Rangers, and almost three times as many agree that laws may have to be ignored to "save the forests." The vast majority simply don't understand the acceptability of tyranny this comic would teach young children. My students are poorly prepared to appreciate, protect, or defend a government based on law. Many would appear to welcome charismatic leaders who would promise to provide environmental security. Fear of an environmental apocalypse has made them ideal subjects for Milgram's experiments. They are well prepared to become functionaries in a government that places "saving the environment" above the law.

Our Institutions Will Save Us From Ourselves

Only the strength of our political institutions and an idealistic faith in "good intentions" stands between radical environmentalism and a collectivist chamber of horrors. The force of "good intentions" is found in the "enlightened teachers" who argue that everyone is capable of becoming virtuous if properly conditioned. Advocates for a "biocentric" utopia argue for remaking people into "ecological selves" who will assume their roles as responsible members of a "community of nature."

I have lived too much history to find comfort in idealistic dreams and faith in virtuous behavior. Too many bruises have taught me that the founding fathers were right about human nature. Accountability for both other people and nature can instead be best achieved by building from the strengths of our traditional political institutions.

Consequently, I am convinced we must stop avoiding the threatening realities of both environmental degradation and social control by a few powerful environmentalists and start talking about how we can make our institutions more accountable for both other people *and* our surroundings—how institutions can motivate people to be more accountable for their effects on others and their environment.

I agree with Vice President Gore on one thing. Nothing is more important for our future political stability and environmental security than the *choices* we make about how to address environmental problems. However, I vehemently disagree that environmental security must become the defining principle of government. Choice of institutional means for solving environmental problems involves little that is new. Many of these choices will repeat the debates that took place when we developed the institutional means for becoming more accountable for the treatment of other people. By addressing the need to become more accountable for how we treat our surroundings, we will be forced to rethink how we can better hold ourselves accountable for meeting the needs of other people.

As I will discuss in the last section of this book, we can best make our institutions more accountable by building upon our historic strengths, not in a fearful rush to protect ourselves from ecological disaster by subordinating liberty to environmental security. Individual liberty, not environmental security, must remain the central organizing principle of government policy toward both people and the environment. The institutional foundations of this nation are freedom of religion, freedom of association, private property, rule of law, and representative government.

Making Peace
With Myself

Why, then, do you look at the speck in your brother's eye and pay no attention to the beam in your own eye? How dare you say to your brother, 'please let me take that speck out of your eye,' when you have a beam in your own eye? You hypocrite! First take the beam out of your own eye, and then you will be able to see clearly to take the speck out of your brother's eye.

Matthew 7:3-5

4

False
Dawn

We shall not cease from exploration
And the end of all our exploring
Will be to arrive where we started
And to know the place for the first time.

T. S. Eliot
"Little Gidding"

Madness is something rare in individuals—but in groups, parties, peoples, ages, it is the rule.

Nietszche

Rippling waters occupy my earliest memories. My first ten years were lived on the bank of a mountain stream. The sound of splashing water accompanied the play of morning light and closed my eyes at night. Our house stood on a high bank bordered by trees. Willows and alders announced the never-ending seasons of life: spring trout and trilliums, summer swimming holes and steak-fries, autumn weenie roasts and the smell of burning leaves, and the silence of winter snow.

The stream was fed by a massive spring high up on the mountain, supplying a constant flow of cold water throughout the hot, dry summers. My father had built a flume to divert a small portion of the stream to a cistern on a pine-covered hill above the house. Green lawns bounded by daisies and damp soil rich with vegetables turned the summer into a childhood paradise. This was my world. This was home.

My childhood dream ended when my parents moved to a larger community to assure a better education for me and my four sisters. They felt a one-room school, with its sixteen children in eight grades, was unlikely to prepare us for life beyond the valley we had come to love.

Sun replaced water in my life. Our new home stood on a hill and looked east over green orchards, oak and grass-covered hills, and distant blue mountains. The morning sun broke the silence of my sleep and dusk sent me scurrying to bed. Even 35 years after leaving this house I still open my eyes with the sun, and still live on a hill where I can see what is going on around me.

Morning light would draw me to the front porch where I would find my mother in her jeans and plaid shirt, her long black hair already wound in a bun on the back of her head, sipping coffee and watching the land emerge from darkness and mist. Over the years we spent countless silent hours in the spell of morning light. These mornings together cemented an unspoken bond. But they also shielded a secret about which I have never spoken and the truths of which I am only now beginning to understand.

After we had lived at the new house for several years my mother woke me early one morning and asked me to join her in the pre-dawn darkness of the front porch. Stars were retreating and the moon had already set in the west. We scanned the eastern horizon and waited expectantly. An expanding ball of light ignited the dark landscape, and just as instantly receded.

In a flash I had seen the orchards in April bloom, cows sleeping in the fields, hills green with spring grass, and pine forests on the mountains. We shivered in the pre-dawn darkness and wordlessly retreated to the warmth of the dining room's pot-bellied stove.

This "false dawn" was more terrifying than I was then ready to admit, but was no mystery. Atmospheric nuclear testing in the Nevada desert 250 miles to the east had been announced in advance. Over the years we rose early several more mornings to watch these "false dawns," but never spoke about the fear they brought to a childhood landscape—fears that would haunt me well into adulthood: fears that would prod me to study nature and to adopt forestry as a profession; fears that would cause me to join mass movements; fears that would lead me to deny the creatures that lurk in shadows of the forestry profession and the environmental movement; and fears that would numb me to the truths of family, friends, neighbors, spiritual soul-mates, and unity with land and country—and death.

The Morning "Nature" Died

"Nature" ended for me on the morning of my first false dawn. In that flash I saw the human hand reach across the land and fasten its grip. The morning spell was broken and the world went limp. Fears of total destruction—an apocalypse—cast a long shadow over my childhood landscape. My adolescent years were occupied by increasing concerns with population growth, resource depletion, pollution, and the need for environmental conservation. My favorite book in high school was John Hershey's *Hiroshima*. The environmental crisis was real! I became a forester, and dedicated my life to land stewardship. Had I grown up today, I would have called myself an "environmentalist."

Living In Two Worlds At Once

Over the years I have gotten to know others who shared my fears of an environmental apocalypse, and have discovered in them the childhood secret I had never admitted to myself: a profound disenchantment with people for causing nature to suffer and an idealized vision of a perfectly harmonious nature. Fear had broken my trust in other people, and I had split off an idealized world of nature from the unreliable world of humanity. Without thinking, I had recovered my childhood dream by adopting a make-believe world of undisturbed nature and looked at people as a nuisance or threat. I lived in two worlds—was occupied by two selves—but didn't know it. Only after I had returned to graduate school to incorporate sociological study in forestry did I become conscious of how widespread the tendency was for people to split humans from nature—although others might trace causes for this split to different sources of fear, pain, and inner struggle.

This split was especially pronounced among the nation's "best and brightest." When I first began blending sociology and forestry at Yale I encountered students or professors who looked at me quizzically when I told them what I was studying. Most responded to my answer with a question: "What do you *do, talk* to trees?" I learned the appropriate response for a Westerner at Yale was to cock my nose at two-o'clock, look indignant, and assert, "I'd rather talk to trees than some people I'd meet!" This would invariably engage them in a conversation about the American fetish with undisturbed nature and romantic distortions of our "nature" as living organisms dependent on the physical and biological world around us.

But only recently have I discovered how severely this split distorts reality by breaking people apart from one another and from the earth on which they depend. This split is a festering wound in our culture, as well as a wound in our selves. It is a source of conflict within ourselves, between ourselves and others, and between us and the biological world around us. The fact that this wound is a *widely shared* distortion of our biological ties to nature does not make it any less serious. It may even make it more of a problem because the rift between people and nature is something that needs to be talked about and understood before it can be healed—before we can be healed. Suffering and conflict will continue until healing begins. Unless we begin talking about the split *within*, we will continue to mistakenly see it in threats from *without*, and seek conflict with others to resolve our inner tensions.

Fantasy As An Escape From Reality

Psychologists have studied mental splitting enough to know a lot about its causes and consequences. People who suffer trauma may escape their painful memories by constructing a fantasy world in which they systematically deny or distort their painful experience and replace it with safer, idealized images. Although denial of painful realities can help people survive, it can also cause individual and collective illness.

Escape from reality can be so pronounced that individuals

severely distort reality. They may see others only as either good or bad—demons or saints—and live troubled lives when these distortions do not match the real nature of others. Or they may be truly "crazy," and escape so far into a world populated by imaginary beings that they behave in a threatening way to themselves or others. We are all familiar with how people distort reality. But we find it far more difficult to see such distortions when they are shared by a large group of people.

Madness Of Crowds

Delusions are just as readily shared by masses of individuals. These collective delusions—shared distortions of reality—can be the basis for mass movements, crowds, riots, and other collective forms of behavior. European sociologists have been more likely to look at the collective delusions as a cause for mass movements, while American sociologists have been more inclined to look at material interests and rational explanations for mass movements and crowd behavior. I have found the European literature more useful for understanding the environmental movement and the radical thinking of powerful environmentalists.[42]

Rapid social and cultural change, together with the individual suffering it may bring, may be one of the primary causes for masses of people to adopt distortions of reality. History is filled with examples of how people have taken on new beliefs when overwhelmed by novel threats or confusion resulting from the collapse of old beliefs and disengagement from social conditions that provided stability.

Mass hysteria was already a subject of study in 1841 when Charles Mackay wrote a fascinating book describing the way people foolishly rushed into speculative economic ventures (*Extraordinary Popular Delusions and the Madness of Crowds*). This book has been republished many times and is still avidly read, although there are sound arguments with many of its generalizations. Bernard Baruch wrote the following comment in a preface to the 1932 edition:

There are other commentators on crowd psychology. The value of this literature lies in its emphasis on forces that are, at all times, functions—and that, at some times, seem to be controlling factors—of national...life.

No prevention is anywhere suggested, but accurate knowledge of and popular recognition of them and their early symptoms should lighten and may even avoid the more harmful of their full effects.

Although there be no scientific cure, yet, as in all primitive unknown (and therefore diabolical) spells, there may be potent incantations. I have always thought that if, in the lamentable era of the "New Economics," culminating in [the crash of] 1929, even in the very presence of dizzily spiraling prices, we had all continuously repeated, "two and two still make four," much of the evil might have been averted.[43]

Scapegoating And Popular Delusions

The current era bears a striking resemblance to earlier eras in which novel threats arose and social and cultural change were rampant. Norman Cohn describes just such behavior in a classic historical treatise entitled *The Pursuit of the Millennium: Revolutionary Messianism in Medieval and Reformation Europe and Its Bearing on Modern Totalitarian Movements*:

But when a situation arose which was not only menacing but went altogether outside the normal run of experience, when people were confronted with hazards which were all the more frightening because they were unfamiliar—at such times a collective flight into the world of phantasies [sic] could occur very easily. And if the threat was sufficiently overwhelming, the disorienting sufficiently widespread and acute, there could arise a mass delusion of the most explosive kind. Thus when the Black Death reached western Europe in 1348 it was at once concluded that some class of people must have introduced into the water-supply a poison concocted of spiders, frogs and lizards—all of them symbols of earth, dirt, and the Devil...the plague continued and people grew more and more bewildered and desperate, suspicion swung now here, now there, lighting successively on the lepers, the poor, the rich, the clergy, before it came finally to rest on the Jews, who whereupon were almost exterminated.[44]

Scapegoating is as least as old as recorded histories and legends. People have always blamed others for problems they could not understand or found especially troubling. Future historians are likely to describe the current hysteria over "Saving the Earth" as such a period. The modern era may turn out to be no different from earlier periods in which rapid social and cultural change and novel threats caused outbreaks of mass hysteria and popular delusions.

Scapegoating And Calvinist Splitting

Cohn's and Baruch's words came immediately to mind when I first saw the blaming of loggers for "raping the land" and "murdering trees." I was even more convinced of the parallels when some university students began to use the term "logger" as a generalized noun referring to "environmental destroyers." While this scapegoating of loggers is now diminishing, and pales in comparison to the genocidal movements of the past, I was struck by the similarity in psychological and social dynamics. I have since witnessed the emerging scapegoating of ranchers for grazing their cows on public lands, and attempts to blame property owners for using lands designated as "wetlands." Attacks by animal rights advocates are equally, if not more, likely to express a Calvinist emphasis on the purity of the elect and impurity of the damned.

Complex environmental problems have been reduced to simple moral choices between good and bad people. Most Americans have been unwilling to face the complex realities surrounding accumulation of greenhouse gases, ozone depletion, chemical pollution, and sustainable management of forests, range lands, and farmlands. A surprising number of these people have preferred to focus their energies on simple moral crusades such as preserving forests, removing cattle from public lands, or denying property rights to private owners in order to capture public benefits at no cost.

Hiding In The Fantasy Of A Magical "Nature"

As a result, the nation's attention has focused heavily on environmental issues that are relatively insignificant when compared with the major ecological problems. The elimination of harvesting on public forests in the Pacific Northwest is an exciting romantic ideal, but it will do very little to "Save the Earth." And its unanticipated consequences may include substantial ecological and social disruption in the Russian

Far East and other regions that may provide substitute sources of wood materials. Removing cattle from public grazing lands and replacing them with buffalo, elk, and deer may satisfy some aesthetic preferences, but may result in far greater ecological disruption when deer and elk populations go through natural cycles of growth and decline. Yet like the celebration of the Yellowstone wildfires of 1988, such ecological disruptions would be seen as "good" because they would be "natural." The pursuit of a magical "nature" is one of the most popular modern delusions.

People appear to be reacting to their fears of an environmental apocalypse by reducing complex social, economic, political, and ecological issues to simple moral choices: good and bad people, pristine and destroyed environments, and peaceful and catastrophic change. This is fantasy, not reality. Calvin's ghost is still at work. These are collective delusions, no different than those that occupied the minds of terrified people throughout history.

The unrealistic simplicity of this fantasy world is epitomized by the media images of "owls versus jobs" and "environmentalists versus loggers." Reduction of complex issues to simple moral choices is a sure sign that modern society is not immune to the sort of popular delusions that have led other people to so tragically persecute others, rush off to war, or throw their money away on fantastic investment schemes. The roots of social conflict and war can be traced to popular fantasies about people who are good or bad because of how they look, what they eat, what they wear, how they make a living, or what god they worship.

Why would our modern, supposedly rational, society retreat to such stereotypes and moral blaming? Have we become so dependent on modern media images that we can only see ourselves as participants in a moral crusade to "Save the Earth?" Have romantic Disney images of nature replaced real experiences?

Yes, we have a tendency to get lost in the fantastic world of media imagery and lose touch with reality. But I believe the not-so-good, old-fashioned tendency to distort realities is at the root of the problem. Modern media simply make it easier for masses of people to participate in widely shared fantasies and to be manipulated by propagandists who spread fear of impending disaster. Fantasies about returning to nature are no different than fantasies about fixing environmental problems by designing better technologies or securing peace by building bigger bombs. They all ignore obdurate realities that will not go away.

Cannot Eat Without Taking From Life

After years of talking with people about our tendency to deny the full reality of nature I have concluded one of the most troubling fantasies is our denial of what I call the "paradox of life"—the biological fact that humans, like all other organisms, cannot live without taking life. Webster's dictionary tells us that a paradox is a statement that seems self-contradictory or absurd, but in reality may express a possible truth. The simple truth of our life as biological organisms is that we must take life from other organisms in order to feed, cloth and house ourselves.

Yet many modern, urban people often deny their own biological reality and live in a fantasy world in which consumer goods appear magically and are not thought to involve taking the life of other organisms—an attitude expressed by some California school children who recently asked: "Do pigs lay bacon?" Similar reasoning causes people to think about building materials coming from the lumber yard, meat from the market, and cars from Tokyo or Detroit. The ultimate source of materials in nature is ignored or denied. Also denied is the service provided by those who grow, harvest, extract, transport, and transform these basic materials—the fact that people producing the resources play the essential role of serving consumers in the resource-dependent cities. Hence, the real "timber-dependent" or "resource-dependent" communities are the cities and suburbs where most of the consumption occurs.

Humans Can Live Sustainably With Nature

The biological truths of human existence have been accepted and embraced by most peoples who have lived with land over long periods of time without disrupting nature to the point where their future sustenance was impaired. These peoples have developed rituals for honoring nature, limiting the amount of life they take, or undertaking conservation practices to perpetuate or restore the productivity of the land upon which they depend. René-Daniel Dubos, in an inspiring essay entitled "Symbiosis Between Earth and Humankind," has told us how peoples around the world have developed conservative land use practices that have enabled them to live for over a thousand years on the same land base:

Nature is like a great river of materials and forces that can be directed in this or that channel by human intervention. Such intervention is justified because the natural channels are not necessarily the most desirable, either for the human species or for other species. It is not true that "nature knows best." It often creates ecosystems that are inefficient, wasteful, and destructive. By using reason and knowledge, we can manipulate the raw stuff of nature and shape it into ecosystems that have qualities not found in wilderness. Many potentialities of the earth become manifest only when they have been brought out by human imagination and toil...the earth is to be seen neither as an ecosystem to be preserved unchanged nor as a quarry to be exploited for selfish and short-range economic reasons, but as a garden to be cultivated for the development of its own potentialities of the human adventure.[45]

Splitting Reality Leads To Guilt About Living

Modern people who reluctantly accept they must take life in order to live often experience a tremendous sense of guilt for what their consumption activities do to other life forms. Guilt may lead them to adopt vegetarian diets, or to live simply. They may deny their guilt and explain how consumption is needed to satisfy present needs. Unresolved guilt may motivate them to join social movements that seek the sort of totalistic solutions to ecological problems I will discuss in the next several chapters. When coupled with an understanding of human population growth, such "guilt about living" can exaggerate disenchantment with people and lead to the devaluation of human life. I have heard radical environmentalists say "There are too many people in the world, so let them (not me!) starve." Fantasies about population reduction can in turn produce further feelings of guilt for secretly wishing destruction on others. All together, such unpleasant feelings of guilt can reinforce the mental splitting in which nature is romanticized and people are blamed for their destructiveness. A downward spiral of guilt and mental splitting is set in motion, and advances under its own power as it feeds on itself. Individuals suffering from this inner struggle are ideal candidates for social control through moral persuasion.

This downward spiral is not endless. Mental splitting can take on a life of its own and be accepted as totally normal. Psychologists have learned that guilt can disappear when mental splitting is cemented in a person's character.[46] Severe distortions of reality can become normal, and people may do horrible things to one another without suffering guilt if they are told their destructive behavior is necessary or virtuous. A part of the self breaks off and sets up an independent existence as a collective self—even something so apparently benign as an "ecological self"— enabling people to guiltlessly kill others in a war or riot, torture prisoners, oppress people they feel threatened by, or exclude those who are thought to be morally inferior because of their religion, race, gender, lifestyle, occupation, or beliefs.

A great many people in American culture do not suffer from this split, either because they have inherited or worked out a harmonious relationship with nature or because their split thinking has taken on an independent life in which they no longer feel the pain. The latter are no longer open to their experience and are numbed by their painful history. The former are often far more practical and willing to change their behavior to help sustain a healthy environment.

There are many of us who have not been able to successfully deny the realities we experience or develop a comfortable relationship to nature. I have found I could not mend my rift with nature until the rift in myself was healed. I now know I can never return to my childhood home, but I can feel at home in a world confused by mass hysteria over a deteriorating environment if I search for truth and ground myself on firm realities.

5

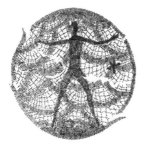

Kill
The
Pig

We are all haunted by echoes from our past—memories we wish
we could forget, regrets for not doing something for a parent or friend,
guilt from hurting someone. As creatures of time, we carry our past
with us. Sometimes it enriches the present by giving us wisdom to make
better choices. Other times it weighs so heavily on us that we grow
numb and, like beasts of burden, plod through each day without thinking or feeling.

I am haunted by Viet Nam. Not the horrid war across the
Pacific, but the war at home; a war in my soul that was ended by a
peace treaty. The violence of Viet Nam is a part of my burden, but it has
also given me eyes to see, words to speak, and truths to proclaim.

Too old for the draft, the 1970s found me in graduate school at
Berkeley, idealistic, and deeply committed to a peaceful end to the war.
Several of my friends and undergraduate classmates had served early in
the conflict. Two had been severely injured and one had been killed. The

human suffering on all sides of this conflict pained me deeply.

One evening after taking part in a demonstration against the war, several friends gathered at my apartment to relax and talk. Dispersing crowds passed under our window, chanting, singing, and waving banners. The drift of the crowd isolated a policeman from the protection of his fellows, and he was pinned against a wall in a parking lot immediately below us. One of my friends noticed what had happened, ran to the window, and totally out of character, urged the crowd to violence by shouting: "Kill the Pig! Kill the Pig!" Fortunately his voice also attracted other police officers, and the surging crowd scattered to escape clubs and tear gas.

As much as I tried to suppress my fears, I realized what I had embraced as "peace and love" also embodied the shadow of something sinister—something so large and uncontrollable that I was swept along with it and couldn't see where I was going or where I had been. I moved to the periphery of the peace movement and grew into a more effective student of sociology—having gained personal experience about how even idealistic and "loving" mass movements are nourished by unconscious images and impulses.

Bearing The Burden Of A Viet Nam Protester

For the rest of my life I will carry the burden of knowing I participated in a movement that professed peace and love and yet inflicted symbolic violence on Viet Nam veterans who, regardless of the ethics of war, deserved the honor and respect of their country. I first began to feel the weight of this burden when my classmates came home, and I vowed I would never again be so naive, and sought a way of transcending such political conflicts. And yet how naive that vow turned out to be!

The Horror Of Meeting My Shadow

I had yet to discover that I carried the potential for violence within me. A strange and undiscovered creature danced in the dark recesses of my inner self, and I had yet to find out what it was and what it wanted. I began to study depth psychology and the social psychology of mass movements to help me reveal this hidden life and understand why I had so easily become a vehicle for evil.

What I discovered first was the *shadow* in myself[47]—an inner, hidden, part of myself that was terrified of light and had escaped detection for over 37 years. What a horrific and disturbing discovery! I was not who I thought I was! Yet, somewhat to my surprise, I was relieved to begin getting to know a part of myself with which I was already very familiar. I slowly began to understand that what I had found most repulsive and threatening in others were often qualities I had refused to see in myself. I cannot tell you how difficult it was to try to see myself as I really was, and to give up the illusion of who I thought I was or would like to be.

We Create Evil To Belong With Others

I have sought understanding in the words of others who had traveled this path before me. Sam Keen stated it clearly when he said:

The most terrible of all moral paradoxes ... is that we create evil out of our highest ideals and most noble aspirations. We so need to be heroic, to be on the side of God, to eliminate evil, to clean up the world, to be victorious over death, that we visit destruction and death on all who stand in the way of our heroic historical destiny. We scapegoat and create absolute enemies, not because we are intrinsically cruel, but because focusing our anger on an outside target, striking at strangers, brings our tribe or nation together and allows us to be part of a close and loving in-group. *We create surplus evil because we need to belong.* [my emphasis][48]

I too have learned that the need to belong, to feel "at home" with others, and to share a sense of place and identity is essential for maintaining a sense of security and peaceful life. Sam Keen points out that we often find belonging by creating a "shared delusional system" in which we and our enemies project onto each other what we cannot accept about our selves:

... our tirades against Soviet state control and lack of individual property reflect an unconscious anger at the real loss of individual freedom under corporate capitalism, and our dependence on the government to care for us from womb to tomb, neither of which fits our frontier image of ourselves as rugged individualists. We officially see their dependence on the state as slavery, and yet we have embraced big government and galloping socialism...They [the Soviets] see us as sanctifying the greed of powerful individuals at the cost of the community ...[49]

Academic Absolutism Is A Home For The Shadow

These were disturbing thoughts. How I longed for my simpler days as a forester, working out my frustrations by crashing through the brush, swinging an ax, and proving my physical stamina, agility, and field skills by measuring myself against the best in the woods. I understood loggers, and I think they understood me, because we shared a passion for hard work, clever problem-solving, physical mastery of our environment, and a good laugh in the face of adversity. Physical struggle had kept my shadow caged. Later, intellectual work introduced my shadow to others, and without knowing it, my shadow found company in academic absolutism and came to own me.

Nothing in academia has been so effective in empowering my shadow and giving it ownership of my self than apocalyptic images (fears of destruction and death) in which an "enemy" is held responsible for threatening the end of the world. For some academics the threatening "other" is often "capitalism," "Christianity," and other "superstitions," or those thought to be responsible for "destroying the environment."

The Apocalypse Provides Security

Perhaps this is why I fastened myself to the words of a Soviet official who in 1987 told American officials, "We are going to do something terrible to you. We are going to take away your enemy. And you are not going to know what to do." Boy was he ever right! But did he know how our search for a new enemy might break us apart and set us at one another's throats? I keep feeling that we are lost without our Soviet demons to give us a sense of national identity and belonging. We did not need to own our selves because we were addicted to fighting an external enemy and could project on the Soviets what we did not want to accept in ourselves.

I have since found a reflection of my shadow in the words of Rene Girard (*Things Hidden Since the Foundations of the World*), who understands how much we grew to need the "bomb" to keep us feeling secure—to replace a God we no longer trusted to provide security. He stated, "Humans have always found peace in the shadow of their idols … This is still true, as humanity looks for peace under the shelter of the ultimate violence … violence prevents violence from breaking out."[50] The threat of violence—the bomb—gave us security and brought us together as a people. You would think Girard had named the "Peacekeeper" missile!

The bomb made a complete apocalyptic idol. The Soviets projected evil on us and we on them. Reciprocal projection of evil was connected with a truly apocalyptic fear of nuclear warfare. We knelt at the feet of violence and let violence become our god. We became addicted to violence. And the threat of mass violence brought us peace.

The ending of the Cold War has produced a moral paradox: as a nation, we found security and belonging by threatening the ultimate violence of nuclear confrontation; the bomb organized our violence by mobilizing our potential for evil and projecting it onto the Soviets. But where will evil be projected when we no longer face a common enemy? Will we invent a replacement for the Soviet demons, or even for the "bomb"? Who are the new candidates for vilification? And what apocalyptic idol will we worship when the mutual violence of nuclear confrontation no longer provides a sense of belonging and security? What will become our new god of violence?

Environmental Crisis Stands In For "The Bomb"

For most of my career I refused to see that the environmental crisis could become this apocalyptic idol—that this false god we would worship would be the threat of environmental destruction. I couldn't see the idolatry of the "environmental crisis" even though I had caught fleeting glimpses of the shadow that Germanic Romanticism cast over America's idealistic conservation tradition (my topic for Chapter 6). I pushed blindly ahead in my efforts to combine sociology with ecology. What did ecology bring?

Seeking Security In The "Web Of Life"

Ecology places emphasis on the *whole* rather than the parts, on the *system* rather than the process by which things change and individuals develop. Ecological thought deals with ideals: harmony, balance, interdependence, unity, totality, and holism are common ideals. Peter Bishop summarized ecological thought as follows: "Life becomes coherence, the Earth a global cell. Humanity is imagined as merely one life form among many, a planetary being inseparably enmeshed in a living web."[51]

Something monumental happened when moon travelers brought back pictures permitting people to see themselves as passengers on a blue and white ball floating through lifeless space. People increasingly began to associate security with protection of a small and fragile planet. Long-enduring images born from looking *out* upon God's universe were replaced with an *inward* view of the earth as a fragile, living whole. For many people, "Earth" replaced God as the source of security.

How appealing I found these ideals before I began to see the shadow lurking beneath the "web of life." My journey down this mistaken path would have ended much sooner had I been able to read Bishop's revealing comments:

> The widespread celebration, even worship, of humanity's
> inescapable participation in the web of life belies...death and
> destruction.... The web is an appropriate symbol for the shadow
> side of the much proclaimed "return to Mother Earth." The web
> is not only a holistic image to be contemplated in wonder but a
> labyrinth down which humanity stumbles after a sense of its
> own identity and security.[52]

Popular Ecology Brings Apocalyptic Idolatry

I now see holism, the web of life, and "Mother Earth" as apocalyptic idols—as replacements for the "bomb"—which was in turn a replacement for God. Holism demands our unqualified devotion to prevent chaos, and threatens destruction if the whole is broken into pieces, if the forest is fragmented, if species go extinct, if bioregional webs of life are torn, if forests —"the lungs of the earth"—are "destroyed."

Did a similar awareness of idolatry lead Bishop to speak of worship and celebration coupled with fantasies of loss, extinction, annihilation, chaos, and fragmentation? The shadow side of holism is apocalyptic violence (fear that the world will end if the biological whole, the web of life, is broken). To worship the god of apocalyptic violence is to be preoccupied with the fragility of wholes and to tremble with fear of chaos if whole ecosystems are fractured by dealing with them as particular places, problems, and people connected by processes. This may be why people are so repulsed by the clearcutting of forests. Clearcutting has come to symbolize apocalyptic violence.

Apocalypse In The Forest

I didn't fully understand the shadow side of holism until I began studying the social impacts resulting from the sudden reduction in Northwest timber harvests thought necessary to protect the northern spotted owl. I looked into the eyes of loggers and other wood products workers and saw their pain and disbelief when they were made scapegoats. They'd been told that individuals who "kill trees," "rape the land," "fragment ecosystems," "murder owls, "and "destroy nature" had little claim to normal fairness and social justice. "Owl murder" had even entered the lexicon of the Sunday comics, a sure sign it was becoming part of popular culture.

I saw this same apocalyptic idolatry expressed in popular culture through films such as *Dances With Wolves* and, especially, *Fern Gully ... The Last Rain Forest*, as well as in Sunday comics, talk shows, elementary education, and Saturday morning cartoons. *Fern Gully* (produced with the cooperation of the Smithsonian Institution) clearly reveals the shadow side of holism. Green and blue-eyed, forest-dwelling fairies are threatened by brown-eyed and darker-skinned loggers who work for a jet-black demon with a terrifying harvesting machine. The movie combined apocalyptic fear, scapegoating, and latent racism in a simplistic moral battle between good and evil.

Every time I witnessed the scapegoating of loggers or other wood products workers I heard an echo from my past chanting "Kill the Pig! Kill the Pig!" answered by a chorus droning in response, "Humans are only one species among others, and do not deserve special consideration—particularly when they destroy ecosystems." This was all too familiar. I discovered the same sinister forces lurking beneath the caring, extended hand of nature preservation as I had found in the Peace Movement.

Apocalyptic violence is a fearsome god, and is so effective at terrifying people that masses voluntarily bow down before this awesome power. Like others who sought ultimate security by appeasing apocalyptic violence, I released the forces of chaos and destruction in my own shadows and risked becoming an agent for evil. A hidden alliance with violence and a hatred for "other" people was the creature I had failed to discover.

Other Creatures Lurk In The Shadows

I have talked about how fantasies of the apocalypse underlay idealistic visions of "whole ecosystems" threatened by chaos. But what about the other side of the conflict? Don't loggers, ranchers, and other producers also have their shadows? Yes, I have also encountered these shadows in myself.

Wood products workers, along with farmers and many other rural peoples have created elaborate, idealistic myths of independence from government and corporate control, individual freedom, private property, and personal trust. These myths have helped protect them from the unacceptable realities of increasing control by multinational corporations and the progressive collectivization of property and social engineering that has accompanied a movement toward centralized governmental control of both the economy and the society. Farmers denied reality and welcomed government payments, and wood products workers welcomed their dependence on the socialized timber enterprises of the U.S. Forest Service and Bureau of Land Management and the subsidies for housing that stimulated the lumber markets.

Shadowy Realities Are Revealed

The loss of a common enemy in the Soviets, increasing concentration of capital, and the "green revolution" have together created a politically explosive situation. Rural people now understand that some of their historic practices were not sustainable, their economies are being controlled by international corporations, their property rights are being transferred to the public at large without compensation, and their government is attempting to socially engineer their lives through education, welfare, work-fare (ecosystem restoration), and retraining and relocation programs. Moreover, some people perceive how many government agencies have stolen their future by exploiting the public scapegoating of loggers to justify massive, "costless" transfers of public assets and private property rights.

Their political crisis is the loss of their way of life and the threats to the American political and economic institutions that built this way of life—private property, democratic governance, economic individualism, and a shared affirmation of family and community ties, individual advancement, and religion. Unlike those who find meaning and security by worshipping a whole threatened by imminent destruction, rural producers are confronting their aggressors. The curtain has been torn. Shadowy realities are being revealed. And just as Vice President Gore prescribed, a civil war over the environment is raging in our culture.

Environment Conflict Is Politically Unstable

These two fearful visions I have described are on a collision course. But unlike the perfect symmetry of delusions and certainties of mutually assured destruction shared by the United States and the Soviet Union, environmental conflict is unstable. One side is paralyzed by fear that the whole will be destroyed and life as we know it will end. It has responded by projecting evil on those who tear the web of life and by projecting good onto a utopian future and/or past golden age of perfect harmony and balance with nature.

Rural producers on the opposite side of the battle line have poked their heads through the tear in the curtain and are mad as hell about what they see happening to America. They are upset by the loss of a way of life, power of big banks and corporations, growth of collectivist government, loss of property rights, and the mean-spirited assault on workers, families, and ways of life. They are willing to compromise to protect the environment, but are demanding a fair, honest, and secure settlement to the conflict.

Vulnerability To Demagogues

Both sides are vulnerable to demagogues who excite passions, identify enemies, and organize the projection of shadows on to these enemies. Believers in the environmental apocalypse have already been captured by academic demagogues who teach "politically correct" thinking about environmental problems and provide the feelings of security and belonging that accompany projection of evil onto common enemies. Mass media, responsible for defining popular culture, have increasingly come to serve as willing instruments for such propaganda.

Rural producers who embody America's traditional values and institutions are especially at risk to demagogues. Most of their elected representatives have abandoned them. They lack effective political voices in government, with the result that extremists are increasingly giving them a needed voice. Various interests urge them to mobilize against internal or external enemies.

Some use fear tactics and stereotyping to mobilize an assault on environmental protection and the "enemies" that promote it. Others, especially some large corporations, urge mobilization against the "enviros" (environmentalists), but have an ambivalent relationship to rural producers; they need their political help in securing raw material, yet they want cheap labor and do not want to be obligated to maintaining rural communities when profits are enhanced by closing an operation.

False Fears And True Problems

Honest communication is urgently needed to bridge this growing rift between these two opposing groups and to avoid the destructive social conflict that will inevitably follow. We need to start talking about how demagogues manipulate our shadows to achieve political domination—how idols of fear and violence are used to mobilize modern masses and transform them into advocates for a centralized cause.

But, most importantly, we need to talk about our fears so that we can distinguish the real apocalypses—such as a the ozone hole or global changes resulting from greenhouse gases—from the imaginary apocalypses of timber harvesting and cattle grazing in North America. These false apocalypses are often cooked up to scare and manipulate people so that they can be controlled by fear, guilt, and shame. Widespread distrust of all scientific warnings about environmental threats is the unfortunate result of such cynical manipulation. This too must be talked about, both among scientists and among everyday citizens.

But the most important talk must happen within ourselves. My participation in the war over the environment can only be ended by ending the war with myself. I must integrate my self. I must own my shadow so that it will not secretly join with others who would like to establish control over me. I must own all of myself if I am to prevent others from exploiting my internal struggle and gaining control over me and using me as an instrument for violence.

6

A
Forester's
Shadow

I did not know the limits of the possible.

Sam Keen
Faces of the Enemy

... The central defect of evil is not the sin but the refusal to acknowledge it.

M. Scott Peck
People of the Lie

Accidents of history often change the course of our lives. How many of us have suddenly departed from the path we were traveling when unforeseen events intervened, giving us new choices or diverting our attention. How soon we forget our travels on the old path and follow new tracks as if they had always been there.

I began to reveal the shadow side of my profession 20 years ago, but I was scared off by the suicide of the professor who was guiding me in this work. For over 20 years I hid behind the guilt of not fully celebrating his life or fulfilling the spirit of freedom he saw in me. I chose to remain comfortable and avoid the shadows toward which he *pointed me.*

I was a Ph.D. student at Berkeley and began investigating the emerging environmental movement under the guidance of a brilliant sociologist at a neighboring university. I was intrigued by the ideology of holism—the commitment young "eco-activists" made to the idea that the whole ecosystem was more important than the parts and that the parts must be subordinated to the whole.

My motivation was personal as well as professional. Something deep in my forester's soul troubled me: a moral imperative emphasizing the perpetuation of the forest as an organic whole had long been part of the forestry profession. The importance of the organic whole had been an unexamined part in my forestry education. Given the predominance of German thought in American forestry, I was vaguely troubled by the association between the German romantic view of nature and the unexamined American commitment to whole ecosystems. I worried over unanswered questions: Who speaks for the whole? Who speaks for the ecosystem? What authority will take control of the whole?

Unlike the romantic tradition shared by many German foresters, American foresters lived in a culture valuing the importance of the individual and abhorring subordination of individuals to the authority of the state. Yet some foresters and many young environmental activists called upon the authority of the state to solve environmental problems. They sought to protect the whole from greedy individuals and short-range thinking. I was vaguely troubled by a shadowy force that was intolerant of individual liberty. This force struggled against my naive attempts to bring it to light.

My sociological guide gave me the courage to begin revealing the hidden powers of holism. He also struggled with hidden powers he could not understand or control—powers that turned out to be far more threatening than the ones I faced. But he chartered a course of study for slowly lifting holism from the dark recesses of a young forester's shadows. I felt the sense of exhilaration that anticipates a breakthrough from murky confusion to the light of understanding.

But it didn't work out that way. My guide succumbed to whatever hidden powers pursued him. It happened not long after I had visited him on the spur of the moment to, in his words, "go and 'mao mao'

*Stewart Udall (visiting at a Democratic Party function) for giving away
so many oil leases on the last day of his appointment as Secretary of
Interior." When Udall proved unassailable, we left the party in my
guide's pick-up truck and raced across town, his reddish-blonde hair
blowing in the wind, his face straining to stay ahead of his pursuer, slid-
ing sideways through stop signs and traffic lights—testing the limits by
literally balancing on the edge, as was his personal style in both sport
and intellectual work.*

*But after I left him he couldn't make himself stop. He kept slid-
ing—through stop signs and traffic lights, down 80 miles of interstate—
not coming to rest until he had closed the door of his mother's garage
and found safety in a cloud of noxious gases.*

*He told me about only a few of his fears, so I'm not sure what
demon pursued him so relentlessly. His intense interest in my search for
the shadow side of holism made me worry that during his many travels
in an underworld of witches and drug-induced hallucinations he had
also encountered holism's hidden powers. I was much too conservative
to join him on trips to this underworld, and worried he had plunged too
far into the depths and had been sucked in by hidden forces. I was afraid
to go to his memorial. I felt he was not such a great guide after all, as he
had failed to show me how to hang on when pulled by strange ideas and
mysterious powers. I retreated from the edge and stayed safe.*

*I gently approached holism in the years following my mentor's
passing—hoping to reveal its powers by watching it at a distance. But I
learned little. I used the classroom as a safe place to expose romantic
ideals of holism by quoting from Franz Heske, the German forester who
visited Yale in 1935 and, in German Forestry, expressed the romantic
view of German forests more clearly than anyone before or since.
Students were often mesmerized by the cadence of his poetic images.
And like many beginning foresters, and today's idealistic earth war-
riors, I was drawn by the appeal of a oneness with nature which I later
discovered had been captured in Franz Heske's descriptions of German
culture and its close affinity with forests:*

> *The culture of the city, with its unceasing human turmoil
> and daily elbow-to-elbow struggle for bread ... moves the little
> Ego into the center and finally causes the whole world to be*

*viewed from this minute observation post. The civilized coun-
tryside, with its flat fields and its innumerable boundaries,
fences, hedges, land boundary stones, is everywhere a reminder
of exclusiveness and segregation, of the Ego and of the micro-
cosm subservient thereto.*

*Not so in the woods. Primordial depths, mysterious mur-
muring and whispering surround the wanderer. Loneliness in
the face of gigantic Nature in which everything is large, everything is complex and yet unified, soon makes the little Ego dis-
solve organically into a new totality. The egoistic soul expands
and becomes like a transparent ball in which the organic
streams of the whole universe flood back and forth. The armor
falls, and man is free.*[53]

*But students also grew fearful and turned away when I showed
how the holistic ideals they found so captivating were instruments of
Nazi propaganda. The hidden force of holism continued to both cast its
spell and resist the light of understanding.*

*I have finally found a secure foothold and can safely return to
the precarious path from which I fled 24 years ago. I can now stare
holism in the face without being hypnotized by its charms and falling
victim to its seductions. What I see is so troubling that I feel compelled
to share it with others. This is for you, Jim.*

Unquestioned Commitment To Holism

A growing commitment to a biological concept of holism is
pervasive in government policy, academic ecology, natural resource
professions, and, especially, the environmental movement. Many
people are embracing "ecosystem management" as the solution to conflicts over managing forests and other wild lands. The problems of preserving individual species of plants or animals, maintaining commodity
production, protecting the quality of water and air, and addressing many
other concerns are thought to be solvable only by focusing on the whole
instead of parts and processes. Rather than building plans to preserve the
northern spotted owl or other species, we are encouraged to build plans
for managing the whole ecosystem.

What is so troubling about the commitment to biological holism is not holism itself. There are many well-thought out philosophies of holism, including much of theology. Instead, there is an amazing lack of critical thinking about the advantages and disadvantages of adopting a holistic view of biological systems. The adoption of holism as a basis for designing public policies is unprofessional and irresponsible unless its hazardous social and political implications are identified, discussed, and avoided.

'Ecosystem' Is A Powerful And Useful Concept

For over 15 years I, too, advocated an ecosystem approach to natural resources management. Some might credit me with playing an important role in promoting the ecosystem approach within forestry. The ecosystem approach provides a powerful mental picture of how nature works. It says that nature can be best understood as interactions between living and non-living things. Plants and animals, soil, water and air, all are hooked together through pathways by which energy is captured from the sun and is passed from one organism to another, and through which materials move in endless cycles.

The ecosystem approach is a road-map to the workings of nature. It tells us how each living thing is part of a complex web of relationships—allowing us to anticipate how particular plants or animals might be affected if we change the fabric of relationships that give them life. The mental picture of the ecosystem helps us understand how to manage *particular things* by making sure they have what they need to live, reproduce, and perpetuate themselves.

The Nonsense Of Managing Ecosystems

The "ecosystem" is a powerful and useful idea. We need the idea of the ecosystem to help us manage our environment, just as we need maps to help us locate things and move about without getting lost. But I am troubled by the recent idea that we should "manage ecosystems." Since ecosystems are mental pictures in our heads, abstractions, not things, we can no more "manage ecosystems" than we can "drive maps." Just as we drive on roads with the aid of a map, we can manage concrete things—soils, water, plants or animals—with the aid of ecosystem models. Ecosystem management mistakenly treats an abstraction—

an idea—as if it were a real thing. This is a serious flaw in rational and scientific thinking. But that is not what troubles me the most. I am haunted by the unspoken, mysterious appeal of holism that calls us to manage the whole ecosystem instead of its parts.

German Origins Of Holism In American Forestry

I will speak directly about the magical, shadowy side of forestry and "biocentric thinking" by examining how holism found its way into a forester's shadows. American forestry was transplanted from Europe, especially from Germany. The romantic visions shared by some German foresters are reflected in the way many American foresters think and feel. Moreover, much of the conservation tradition which has found expression in the modern environmental movement stemmed from early American foresters who brought ideas from Europe. Holism was a central theme in German forestry, and as such, was unthinkingly adopted and became part of the American forestry and conservation tradition.

When I received my forestry education at Berkeley there was no clear definition of the whole forest as is now found in the ecosystem concept. Holism entered my thoughts unannounced through the "doctrine of the long run." Foresters think in terms of centuries, not years, or even decades. Foresters have always struggled with the problem of ensuring social conduct that would perpetuate forests for generations to come. Given American forestry's European origins, it was assumed that the central authority of the state had to speak in the interests of the long-run. The only question was whether the state would compel people to conform by passing laws or whether it could shape people's behavior through education or persuasion. I did not stop to think about who should be empowered to speak for the long-run—who would advocate the larger whole—or whether there were ways of ensuring the perpetuation of forests other than by coercion or persuasion by centralized authority.

I was instead enchanted by forestry's mission of working in the interests of posterity. Nothing in forestry more clearly symbolizes this commitment to posterity than the ideal of creating and perpetuating an even distribution of age classes—a commitment to sustaining the yield of goods and services over the long run. This ideal subordinates individual gain to public welfare so that the future can be protected from short-range profit-taking. But what I failed to see when I entered forestry, and could not look in the eye until recently, was the underlying myth that motivated this commitment. This may be the founding myth of the romantic strain in German forestry.

The Folk Organism As The Whole

A myth is a story about supernatural events, gods, and mysterious powers that are presumed to be active in the world. Myths may be stories about mysterious personages. We tend to look upon myths as superstitions and untrue stories believed by irrational people who lived long ago. But what we fail to see is how current myths give people motivation, purpose and direction. Living myths (those in which people actively believe) project an aura of danger, sacredness, or power that protects them from being questioned. Some would say we may not be able to live without myths. So it is terribly important to know about living myths that point us toward paths and propel our movement. But to question myths is to court danger by disturbing those who believe them and stirring up our own superstitious fears of treading on sacred ground.

I was at first terrified to look holism in the eye. Something powerful wrestled within me and sent me scurrying away to safety. Two things frightened me. Holism was a powerful idea and carried an aura of sacred mysteries usually associated with spirits or gods. Holism also was too closely tied to the romantic ideology that had given rise to Nazi Germany. I was fighting with myself—fighting to free myself from the grip of a powerful myth I believed without *knowing* it. By putting words to this myth Franz Heske helped me see it:

If the principle of unrestricted economic egoism, i.e., *laissez faire*, had dominated forestry no such even distribution of the age-classes would be present. This age-class structure shows that the past generations did not merely look after their own immediate advantage, but also bore in mind the interests of posterity. They considered themselves as *only the temporary members of a perpetual folk organism.* [my emphasis][54]

People Are Subordinate Parts of an Organism

Along with many of his contemporaries, Heske saw society as a living organism in which individuals were subordinate parts. And people who cared for forests were soldiers (just look at the military uniforms of German foresters and their apostles around the world) working to perpetuate a society by protecting its culture:

> The cultural importance of the German forest population as a fixed pole with an instinctively sure inner compass cannot be valued highly enough in a time of general shifting of all values. An intimate connection with the soil, such as results especially from ownership of a long-lived forest with its sustained management extending through many generations, is the strongest bulwark against social chaos. Woe betide, if the forests were to be partitioned! A mass of culture and perhaps its very backbone would be destroyed.[55]

Questioning The Power Of The Whole

This is *powerful* mythology. People and forests both participate as subordinate members of an organic whole, and the fracturing of this whole would produce chaos. I began to understand why I had felt so afraid whenever I tried to look at holism. I now know why I felt like I was threatening chaos by questioning the whole. By analyzing this mystery I was "fracturing the forest." I was fracturing the organismic unity of society. I was undermining the cornerstone of my sense of security as a forester! I was challenging the founding myth of my profession. "Woe betide, if the forests were to be ..." fragmented, social chaos would descend upon us and tear us apart.

Like all good soldiers, I stood straight and stayed in line until the war over the environment broke out. I was in Washington, D.C. presenting results of my assessment of the social and cultural consequences of plans to conserve the northern spotted owl. Several members of the media at a national news conference were visibly disturbed by my predictions of substantial human suffering among woods workers. I had touched a very sensitive nerve, and drew hostile questioning designed to embarrass me on national TV. Their questioning told me I had tread on somebody's sacred ground.

Americans Adopt The Myth Of The Eternal Forest

While staring at the ceiling in my hotel room that night I realized I had threatened people who find security in the image of nature as a stable organic whole undisturbed by human intervention. I had dared to question the sacredness of the spotted owl as an icon representing the stability and holistic unity of old-growth forests.

The underlying meaning of the media's questions hit me squarely in the face: "Why even talk about a few loggers and mill-workers when the unity of the whole is threatened? They simply have to be sacrificed to preserve the whole. And besides, they are only a few powerless people who work with their hands and do not understand the laws of nature." Heske's words streamed through my mind as I came to realize how pervasively the romantic myth of the forest had spread from a few American foresters and environmentalists to the population at large. Forests had become a central myth in American culture! Now the forests were haunted by Calvin's ghost, and those who questioned the holistic unity of the forests placed themselves at great risk of moral condemnation!

Murdering Life Itself

Like Heske's Germans, Americans were finding a sense of spiritual continuity and security by protecting and perpetuating forests: "An intimate connection with the soil, such as results especially from ownership of a long-lived forests ... is the strongest bulwark against social chaos." But Americans were different. The myth of the eternal German forests had roots in practical concerns with providing future generations with wood materials, but Americans wanted to close the forest to most of humanity and to exclude commercial uses.

When thinking about why there was such anger directed at loggers I also recalled Robert Lifton's words about the connection between continuity of life and victimization:

> ... victimization is the creation of a death-tainted group (of victims) against which others (victimizers) can contrast their claim to immortality. Victimizers actually experience a threat to the life of their own group around which they justify their actions.[56]

Cutting trees was interrupting the continuity of life symbolized by the "Ancient Forest." I was witnessing the creation of a living myth. Forests had become a religious icon providing a sense of security for people terrified by an increasingly uncertain world. Fear and hatred of humanity, coupled with the sense of spiritual security provided by undisturbed forest created a uniquely American expression of holism. And those who were seen as cutting these trees shook the foundations upon which others were stabilizing their lives. Loggers were not simply misguided, they deserved to suffer! After all, they were murdering life itself!

Freed By Fear To Study Holism

I entertained myself through sleepless hours by writing scripts for street theater in which displaced woods workers would don prison uniforms and perform in front of the White House. But the next morning I left my fantasies on the ceiling above my bed and rushed home to again take up the path toward revealing the hidden powers of holism. A fear greater than public shame, and a faith in the ultimate wisdom of the American people and resiliency of their political institutions, freed me to again try staring holism in the face.

I am not alone in examining holism. Considerable work has already been completed and more is under way. Two works are especially important. Jan Christian Smuts, the South African politician and philosopher, was the primary philosopher of holism. His book, *Holism and Evolution*, was for years a standard reference. Anna Bramwell has written extensively about holism in a comprehensive socio-political history of ecology entitled *Ecology in the 20th Century: A History.*

As with all such exploratory scholarship, there are problems of over-generalizing from limited case studies or from one author's interpretations. Regardless of these problems, several common themes have emerged for me. The ideas embodied in these themes are especially relevant to current attempts to manage whole ecosystems.

Reborn In The Whole

Holism is almost always associated with a romantic view of

nature in which boundaries are blurred and the individual merges with the whole. This image of being "re-born" through merging with the whole is common throughout the literature on holism. Nature is infused with quasi-religious meanings of cleanliness, harmony and perfection. Romantic images in German culture, from which much of modern American forestry emerged, revolved around a vision of society and nature as an organic whole. Heske expressed these images in his descriptions of small egos blending with the forest in an organic unity.[57] Society was an organism in which individuals depended on the whole— in which the forest, the people (the *volk*), and the state were one and the same. The organism-like reality of the nation state was a powerful myth for controlling diverse and conflicting individuals and groups, and gave them "new lives" and hope for the future. There was a close connection between holism and the "new person" created by the state. The nation state was the creator of the "new person." This is clearly portrayed in the 1937 Nazi propaganda film, *Ewiger Wald*, (Eternal Forest). State film-makers used images of forests to symbolize the holistic unity of the German people and their forests, and emphasized racial purity, forest purity, creation of living-space, personal transformation, and biological/racial continuity as the basis for spiritual continuity.

Regulated By The Whole

While studying "ecosystemology" as an idealistic graduate student at Berkeley, Smuts made great sense to me. Smuts emphasized the workings of natural law to describe how evolution had led to centralized functions which exercise control over parts. Smuts ended a list of how "Holism ... expresses itself and creates wholes." with three revealing statements:

1. "The central control becomes conscious and culminates in Personality; at the same time it emerges in more composite holistic groups in Society.
2. "In human associations this central control becomes super-individual in the State and similar group organizations.
3. "Finally, there emerge the ideal wholes, or Holistic Ideals, or Absolute Values, disengaged and set free from human personality, operating as creative factors on their own

account in the upbuilding of the spiritual world. Such are the Ideals of Truth, Beauty, and Goodness, which lay the foundations of a *new order in the universe.*" [my emphasis][58]

In short, Smuts asserted that centralized forces or principles coordinate and regulate the parts that make up wholes: individuals are regulated by their personalities, persons are regulated by groups, persons and groups are regulated by the state, and the state is regulated by "absolute values." This all fits perfectly with the image of a hierarchy controlled by philosopher-kings who "teach" the masses to do whatever is in the interests of the whole.

The Whole Takes The Place Of God

Like most holists, Smuts believed that the "… ideals of Well-being, of Truth, Beauty, and Goodness are firmly grounded in the nature of things, and will not eventually be endangered or lost."[59] The good, the ethical, was to be found by understanding natural processes. Nature's laws of Holism (Smuts always capitalized holism) had replaced God, and God was reduced to the "mediator of action between the mind and the body … between the physical and the spiritual."[60] To live by these "absolute values," all people had to do was to discover these natural laws and follow them. This fit very nicely with the natural laws of Social Darwinism that had been so popular in Germany a half century ago.

Guided By The Laws Of Nature

I first began to see the shadow side of holism in forestry when I read, and then re-read, Heske. Heske reflected holism by emphasizing the unifying power of natural law when he remarked that foresters "gradually realized that forestry which is to be truly profitable in the long run can never contravene natural laws but must be based upon them."[61] He condemned forestry monoculture and favored indigenously German trees. The contemporary German forest historian, Hans Rubner, once told me this romantic theme reached a point in the 1930s where German foresters were proposing to rid their nation of all "non-Aryan" trees. Heske was less strident and praised the Dauerwald (continuous forest) movement by emphasizing "the fact that the forest is not merely an aggregation of individual trees, but is an integrated, organic entity, comprising all the innumerable living organisms that exist...."[62]

Bramwell, among many other scholars, also helped me see into the shadows of my profession by emphasizing the importance to holism of the romantic image of "natural law" which subordinates humans to the "moral imperatives" of natural processes. But the key theme around which Bramwell organized her book sent blinding spotlights to the deepest recesses of my shadows: "why [as an earlier scholar had asked] the Nazis were 'the first radical environmentalists in charge of state.'"[63]

A Common Shadow

Holism is common ground for most biocentric environmentalists and many foresters. The adoption of "ecosystem management" by the U.S. Forest Service and other land management agencies is to a large extent built on this common ground. I cannot begin to predict how a shared commitment to holism will be expressed. While it would be totally irresponsible for me to suggest that there are elements of fascism in these commitments to holism, there is a serious risk of encountering such dangers in the future if the shadow-side of holism is not exposed and discussed. The rigid adherence to "laws of nature" so typical of many foresters could be expressed in a commitment to "ecosystem management" using new technologies of "adaptive management" to bring humans into conformity with the "laws of ecosystems." An alternative expression might be the celebration of a mythical "wildness" shared by biocentric environmentalists who favor a return to a simpler time (while ignoring the environmental devastation that generally accompanied the struggles to extract a living from raw nature). This would require reduction of population densities and a commitment to "live in harmony with nature" in semi-anarchic "bioregional communities." Removal (through "voluntary" migration) of resource producers from the rural landscape may be the first step toward implementing this expression. Both of these possible cultural developments ignore the shadowy realities of holism and the coercion it embodies. They also ignore an aversion to people that has long inhabited the shadows of the forestry profession.

Separating Ecosystems And Holism

My personal struggles with forestry's shadow are not simply private events. Bringing these shadows to the surface where they can be examined and discussed may be one of our most important tasks in developing better ways for protecting and managing our natural environment. Many foresters have silently struggled with these same questions and have made clear choices to adopt principles freeing them from confusing historical baggage. I think these hidden realities must be brought to light. Unless we can free ourselves from the seductions of holism and the hidden forces it brings into our lives, we won't be able to successfully use the ecosystem concept. The ecosystem concept is so important to helping us learn how to live with nature that we must expose its secret marriage to holism, and, if necessary, demand an annulment.

Holism and ecosystem science are not necessarily compatible. The ecosystem concept can be productively joined to decentralized and loosely jointed forms of social and political organization. These alternatives are far more acceptable to Americans and fit better with our democratic institutions. I will emphasize some of these alternatives to holism in the last five chapters of this book.

7

Green
Guilt

... There is no emotional bondage greater than that of a man whose entire guilt potential... has become the property of ideological totalists.

Robert Jay Lifton
Thought Reform and the Psychology of Totalism

As a child, I had always seen my father as a tower of physical and spiritual strength. The muscles in his arms and chest rippled as he pitched hay onto the trailing wagon, and I guided the tractor through the fields on long, hot summer days. I saw him almost effortlessly lift 200-pound rocks over his head as he built stone walls and fireplaces for the country retreats of city folks. He didn't simply live on the land; I was sure he had grown from it. I could tell because his skin was baked as red as the hard clay on the hills, and his hair fluttered on his frame like the leaves on the oaks. He was like the trees themselves, with his feet planted firmly on the hard earth and his head brushing against the world of spirit.

We all knew he knew God in a special way. He was a seeker, and looked for spiritual truths in all he read or did. Others sensed he knew God, so he was sought after as a counselor for the dispossessed, home-

less, and lonely. He was a vehicle for truths few others understood or could talk about. I felt the movements of the Holy Spirit more often in his presence than with anyone else I have known. As his son, I felt small and vulnerable in the presence of something so large and powerful.

Then one day in the early 1950s I watched him shrink and become like all the other men I knew. He slouched around the house and his spirit retreated into the back of his eyes and didn't come out for a long time. He acted like he had seen a ghost and was uncharacteristically irritable, serious, and afraid.

Three men had driven up to the ranch house in an unmarked county car. They wore business suits and covered their heads with wide-brimmed hats. They took my father into a private room and talked all day. Then they returned another day and repeated this strange meeting. I could tell something bad was happening.

When they had left my mother told us that the County Sheriff had brought FBI investigators to interrogate my father about some friends he had when he came to the county in the 1930s. One of his friends had a sister who was thought to be a communist, and the FBI was talking with everyone who had known her to help identify all the people in her circle who might have been communists during the 1930s.

Only later did I learn about Joe McCarthy and understand how severely this interrogation must has shaken my father's self-esteem and faith in the U.S. government. Why would the FBI question the integrity of a patriotic rural Republican known for his dedication to the teachings of Christ? This all made no sense to a child, but the effect it had on my father certainly left me with a lasting memory of fear. Having watched him shrink, I later understand why the only time I ever heard him raise his voice in anger and swear at another person was when he chastised the chairman of the county Republican committee for violating the trust of his supporters and supporting the "red-baiting" Richard Nixon at the 1952 Republican Convention.

Without knowing it at the time, I had learned something very important about powerful forces that can cause strong people to shrink and grow quiet. Over the years I have watched a lot of strong people shrink. Yet, unlike my father, many failed to regain their stature. Although I never again saw my father as an indestructible outgrowth of nature, his spirit returned with renewed strength and I never saw it retreat until I sat by his side and watched it leave his 97-year-old body.

Only recently have I learned about the forces that had such a powerful effect on my father. Without knowing what I was looking at, I had caught glimpses of these forces when I visited Chinese forestry institutions in 1983. A pervasive power was pressing in on me and forcing me to grow smaller. The communist "state" was everywhere and controlled everything. I began to feel claustrophobic in hotel rooms and hungered for the open spaces of home. Never have I been so grateful to return home to a nation founded on personal freedom. I understood why people knelt and kissed their home soil upon returning from a foreign journey. In a small way, I had discovered what had made my father shrink.

Living A Lie

I again felt these same pressures pushing in on me when I began to talk publicly about the way loggers and other woods workers were being affected by efforts to save old-growth forests and spotted owls. I never argued in favor of cutting old-growth trees; all I did was to affirm the humanity of those who had and did. Most often, students, colleagues, and personal friends grew quiet, as if I had said something which stamped me as being a bit strange, or out of touch with reality. Others would become very emotional and argue with me. One dinner party with close friends erupted into such an emotional argument that both couples phoned the next day to restore the bonds of friendship, but avoided the topic that had caused the outburst. I learned it was much easier to avoid this topic in daily conversations and never let most people know what I thought. I began to live a lie. I caved in to social pressure and kept the truth tightly locked away in a safe place.

But living a lie did not last long. I was growing too small to contain all the anger and frustration I had bottled up. Innocent people in rural wood-producing communities were suffering intensely, and I went on living with others as if the truths of this suffering could be shut off in a hidden part of my professional life—turned on when I was with people who already knew what was going on and turned off when it was not "politically correct" to talk about it. This split world was familiar terrain.

I had seen it before. But where? Where else had I felt myself shrink like this?

Rediscovering Totalism

I was drawn to my bookshelf and soon found myself re-reading Robert Lifton's *Thought Reform and the Psychology of Totalism: A Study of "Brainwashing" in China.* While it was ridiculous to think that the uniformity of thought and social pressure I had experienced was a form of "brainwashing," I couldn't seem to refute the idea that it was a form of totalism. I shouldn't have had to read Robert Lifton to figure this out. To be told that my ideas were not "ecologically correct" or "politically correct" should have been a sure sign of totalistic thinking. While most of my friends and professional associates are too liberal-minded to overtly mark such boundaries for permissible thought, naive undergraduate students were not. They gave me all the clues I needed to see why I was "politically incorrect" for attempting to humanize those who "destroyed the earth."

Lifton placed the Chinese "cultural revolution" in the broader context of "ideological totalism," including the Inquisition, other forms of religious fanaticism, McCarthyism, and extreme political movements when he said:

In discussing tendencies toward individual totalism within my subjects, I made it clear that these were a matter of degree, and that some potential for this form of all-or-nothing emotional alignment exists with everyone. Similarly, any ideology—that is, any set of emotionally charged convictions about his relationships to the natural or supernatural world—may be carried by its adherents in a totalist direction...*And where totalism exists, a religion, a political movement, or even a scientific organization becomes little more than an exclusive cult.* [my emphasis][64]

What Is Totalism?

It is important to understand what Lifton means by "ideological totalism," which I will shorten to "totalism" for the remainder of this discussion. He was talking about an "extreme meeting ground between people and ideas"—people who had become so extreme in their emotional commitment that they couldn't entertain any other point of view. They adopted ideas so extreme that there was no tolerance for competing, or even complementary, views. Totalism is a form of zealotry. It only allows one view of the world, and it excludes or attempts to transform those who do not view the world the "correct" way. Environmental totalism is alive and well in the United States, especially among many school teachers, journalists, politicians, the clergy, and many faculty and students in our universities.

The Hidden Powers of Guilt and Shame

How is it that so many people have shrunk when faced with environmental totalism? Why did I so easily lock my visions of the truth away for safe keeping? There are no easy answers to these questions, but Lifton gives us some insights that can help each of us understand how totalism works. Totalism gets its power from guilt and shame. In China, only when guilt and shame failed to bring about "correct" thinking did totalism revert to violence. By thinking that totalism draws its power from the "barrel of a gun," we fail to see its most effective tools. Guilt and shame can make people shrink and grow quiet, but violence may be needed to cause the most recalcitrant to conform. For this reason, guilt and shame are sufficient forces for transforming most people, and selective violence is reserved for the few who resist such emotional manipulation.

Fear Prepares The Way For Guilt And Shame

Threats of violence have uses other than selective control over those who fail to conform. Generalized fear of violence makes people feel scared and insecure. It makes them doubt themselves, makes them see themselves as "bad," and makes what they think, feel, or do for a living seem wrong. It loosens their ties to daily routines and beliefs. It causes them to begin searching for clues to directions in which they can find stability or order—even to the point of seeking a "new self." This is why revolutionaries and terrorists bomb public places. It shakes up the population and loosens emotional ties to established institutions. But it is also why those who have declared war on our "dysfunctional

civilization" stir up internal conflict and fear of an environmental apocalypse.

How often have my children returned home from school in fear,
having been told by their teacher that "We won't be able to breathe and
the earth will die if we cut down our forests." Fear clouds understanding that new forests will grow to provide the necessary oxygen. Fearful
people will turn away from established ways of doing things and look
for security and belonging in a new order. But fear also prepares the way
for the use of guilt and shame to construct and enforce a new order.

Fear of an environmental apocalypse is a powerful instrument
for creating masses by dislodging individuals from their traditional religious beliefs and commitments to family, community and voluntary
associations. "Ecological guilt," that which I will call "green guilt," is a
powerful tool for shaping these masses and bringing about uniformity of
thought and intensity of feeling. By creating the mentality of a crowd
where everyone is excited by a shared passion, totalism can energize
people for action and give them the feeling of confidence that comes
with the exhilaration of a mass moving in unison to solve pressing problems. However productive these polarized emotional tendencies may
sometimes be (winning a war, for example), and however much good
can sometimes be accomplished by the mass movements that embody
these polarized emotions, they also pose what Lifton calls "the gravest
of human threats."[65]

How Guilt And Shame "Engineer" Emotional Bondage
Dividing The World Into The Pure And Impure

As I have recently experienced, people grow small when they
are told they are not "OK," are robbed of their selves, and are placed in
emotional bondage. Control is established over individuals by portray-
ing human experience as sharply divided into good and evil, clean and
dirty, or pure and impure, and then telling them that they will feel good,
clean and pure only when they adopt the "correct" ideas, feelings, and
actions. Idealists, including many young people, are especially vulnerable to this simplistic appeal. Absolute purity is assumed to be attain-
able, particularly if all impurities are eliminated. Once a commitment to
ridding the world of impurity has been established, almost any action,

no matter how cruel, can be justified as moral behavior.

The Guilt And Shame Of Imperfections
What is ignored, or actively suppressed, is the reality that no person, group, or government is capable of attaining such perfection. When an all-out war on impurity causes people to place such unattainable expectations on themselves, people can readily be manipulated by playing on their sense of guilt and shame. When guilt is manipulated, individuals develop an expectation that they should be punished for imposing harm on other people or nature. And shame, as its companion, elicits an expectation of public humiliation and ostracism for failing to meet the prevailing standards for good and clean living.

No wonder my father, as a seeker for spiritual purity, was so troubled by questions about his possible association with "evil, dirty, and dangerous communists." For related reasons, I too have been vulnerable to being manipulated by guilt and shame. My father's faith was sufficiently strong to fight off this attempt to shrink him and place him in bondage to totalism. And I did not understand the importance of nurturing my inner spiritual resources until I felt the shackles of environmental totalism cramping me.

Pilgrimages To Forests As A Cleansing Ritual
No wonder so many idealistic students tell me about their need to escape to undisturbed forests to "cleanse themselves." They have been made to feel guilty for living, and are ashamed to be consuming goods in affluent urban environments. Our common failure to resolve the "paradox of life" has permitted the spread of a totalist ideology that has exploited "green guilt"—guilt rooted in the "imperfections" of organismic needs for food, shelter, transportation, and health care. (In the next chapter, while discussing the emerging religion of nature, I will focus on the paradox implied by the question: Is to live a sin?)

The popular definition of "forest" has changed to accommodate these cleansing rituals. To a forester, a forest is an aggregation of trees occupying a space. But to the many members of the public and the urban press, a "forest" has become an extensive whole consisting of undisturbed land and trees. A "forest" is distinguished from a "tree farm" where trees are planted, cultured, and eventually harvested. The first is pure and the second is impure, or utilitarian.

I didn't understand this distinction until forestry students (who I thought should know better) began to define "deforestation" (technically defined as the removal of forest cover and replacement by other vegetation or bare ground) as the cutting of old-growth (undisturbed) forests. Despite my protests and appeal to professional source materials, they insisted that "forests" were "destroyed" by harvesting. By popular standards of purity and impurity, they were "correct" in their definition. The totalists had done their work, and the forest sociology professor returned to reading Lifton.

Relieving Guilt By Repenting Or Facing Extinction

The burden of guilt accumulated from inevitable imperfections causes people to be vulnerable to manipulation, especially when the threat of extinction is coupled to the gift of forgiveness. Lifton tell us that totalists have appointed themselves as the *ultimate* judges of good and evil. They have assumed the position of deciding what we should do and not do. Along with this, they appoint themselves as judges of who should have the right to exist and who should not possess this right. This makes ostracism more deadly than public shame, since people are threatened with being deprived of their right to existence within a group or society. The threat of ostracism generally causes people to adopt a "new self" in order to fit in.

The fear of being categorized as a "non-person" is a powerful instrument for ensuring conformity. To be disconfirmed as a person, to be annihilated, is a terrifying experience all people seek to avoid. Widespread vilification of loggers and other wood products workers early in the "war over the environment" effectively denied them their right to exist as equal members of society—a social trauma from which they have not yet recovered, and may never recover (See Chapter 9).

Few people dared risk personal annihilation by associating themselves with "loggers." Leading advocates for old-growth preservation even used this threat to try to cause a rift between the executives for large private industrial holdings and "loggers" dependent on federal timber supplies. They urged industrialists to join them in excluding "loggers" who were cutting trees in national forests.

Forgiveness for living an impure life can ensure conformity when coupled with a threat of annihilation. Repentance relieves guilt and elicits commitment to "correct" behaviors. Current attempts to "transform" rural wood producing communities by encouraging dislocated workers to participate in the "restoration" of lands they had earlier "destroyed" is a powerful gesture of forgiveness—especially coming on the heels of moral exclusion (annihilation of their "old self"). The opportunity to be "re-born" as valued members of society is a temptation many dislocated workers will not be able to resist—especially when coupled with the opportunity to have a job in the woods. But to enjoy re-birth, they must experience "green guilt" and accept their recent suffering as deserved punishment for "sins against nature."

Relieving Guilt By Attacking Others

The flip-side of guilt is an exaggeration of certain personal virtues. This extreme polarization of the way individuals view themselves leads them to see their impurities as the product of threatening outside influences. They relieve the burden of guilt associated with these impurities by denouncing others who do not conform to the "correct" way of thinking. The greater their guilt, the greater their hatred, and the more intense their attacks on others. Lifton summarizes this as follows:

> In this manner, the universal psychological tendency toward "projection" is nourished and institutionalized, leading to mass hatreds, purges of heretics, and to political and *religious holy wars.* [my emphasis][66]

My father experienced the passion of these irrational denunciations during the 1950s. Yet I learned little, and did not discover the destructive potential of such hatred until I found myself in the middle of the religious war over the environment.

Manipulation By Creating A Mystical Environment

There is amazing predictability to the behavior and emotions of those who feel they are spontaneously following their "natural" inclinations, and who feel they are in "in touch with the rhythms of the

Earth." I first noticed such highly patterned or planned "spontaneity" among "hippies" during the 1960s and 70s, when "individual expression" followed a tightly defined script. What Lifton calls "planned spontaneity" can be simple group conformity, as it was in the 1960s, or it can be the product of environmental preservationists who establish scripts that elicit predictable patterns of behavior and emotion through moral persuasion. The carefully orchestrated "planned spontaneity" of the crowd at the Saturday football game is just one example of how all people get caught up in the spirit of a crowd and feel transported by a near-mystical experience of unity and excitement, without realizing how their behavior has been orchestrated.

The mystique associated with a centralized aspiration can effectively channel individual behavior, especially if people feel they are directed by the elect—the chosen few who have a moral imperative to "Save the Earth." But when individuals lose trust, such higher purposes cannot sustain individual commitment. Trust is replaced by the "psychology of the pawn," and people feel themselves caught in a web of powerful forces they cannot understand or control. Insecurity and fear leads them to lose confidence in themselves and to subordinate themselves to those in control. Like Milgam's subjects, they adopt the language and follow the directives of experts.

Control Over Sharing And Producing Information

Control over human information is essential for guilt and shame to be used effectively in fully engineering conformity to "correct" behavior. A simplistic view of the world as a struggle between good and evil, purity and impurity, cannot be sustained when people are given a wide range of information, including information about how elites play the game of moral persuasion and use totalism to get their way.

People who justify control over information assume omniscience and are convinced that reality is their exclusive possession. They know the ultimate "truth" and feel it is their duty to make sure others also know this "truth" and abide by it. Lifton tells us that "In order to be engineers of the human soul, they must first bring it [truth] under full observation and control."[67]

This is what had made McCarthyism so hostile and suffocating to my father. As a seeker for truth, he knew that truth was elusive, and that doubt, personal searching, and ambiguity created the need to continually combine external information with internal reflection. He played by the old rules and knew the ultimate form of arrogance was to adopt a pretense to omniscience. In this role, people assumed the position of God and looked down upon their fellows as unknowing subjects requiring forceful guidance toward the "truth" and the virtuous life it brings.

Guilt-Ridden "Group Think" In Our Free Press

In the United States we are fortunate to have a constitution that protects us from total control of information by a ruling elite. However, we have less protection from the "group think" of individuals in the mass media who seek to control or distort information because it helps them manage their "green guilt" or contribute to control of the masses through moral persuasion. Guilty and fearful journalists in the mass media are at the center of an attempt to control communications by reporting only what fits the "correct" view.

I certainly am not the first to observe how the major national television networks and most of the daily urban newspapers and weekly news magazines (with the exception of publications such as the *Wall Street Journal* and *Christian Science Monitor*) have slanted reporting toward apocalyptic messages and avoided balanced reporting of environmental information. Many of the realities have been virtually ignored: resource consumption driven by population growth, complex structure of land ownership and forest product industries, inspirational stories about the successes of citizen initiatives and private incentives, and the suffering of wood products workers, ranchers, and others dislocated by efforts to preserve the environment. The media have not provided American citizens with a balance of facts and viewpoints on these complex issues. They have instead presented sensationalized battles between good and evil forces such as greedy loggers and fragile owls. Environmental zealotry is alive and well among journalists.

Powerful leaders in the environmental movement feed mass media with "news" as part of their efforts to maintain membership and voluntary contributions. The media thereby serve as important instruments for a conflict whose objective is social control through moral persuasion. But some journalists have not been content to simply adopt available information to "make the news." Tom Brokow recently apologized for NBC's use of video footage showing forest and stream destruction attributed to the Clearwater National Forest because the film was shot at another location where disrupted streams could be found. (This abuse of the public trust was minor compared to the earlier use of incendiary devises to stage a fiery crash so that NBC could show the hazards of a General Motors vehicle.)

How can we expect the American public to have a balanced view on environmental issues when the media have slanted their reporting and stirred up public fears instead of using unbiased reporting to raise legitimate concerns and multi-faceted understanding? Major media appear to be willing instruments of thought control and propaganda. The game of moral persuasion has, for many journalists, replaced the old rules of informing the citizenry so it can be prepared to make wise social choices. However, very importantly, Americans still have the freedom to initiate and support competing communication channels in which guilt and shame are not the business of journalists.

"Sacred Science"

I learned early as a graduate student how tempting it is to maintain an aura of sacredness around my favored theories and hypotheses. These were *my* creations and I wanted to nurture them and protect them from external threats. It took me several years to adopt a self-critical stance in which I could subject my scientific claims to rigorous criticism. Hence, it was not difficult for me to understand how environmental totalism protects basic assumptions and dogmas from critical questioning.

Green guilt tends to have the same effects on the production of scientific information as it has on news-making. Many researchers feel the pressure to produce facts that will be acceptable to their peers or more general audiences. A soon-to-be-classic example of such distortion of facts is the early research on the northern spotted owl.

Researchers hypothesized that owls require old-growth forests and could not survive without them. This hypothesis was tested by looking for owls in old growth, and, not surprisingly, finding them. Only later did some researchers begin to look systematically for owls in forests that had already been harvested, including second growth forests. Had researchers been interested in challenging the hypothesis, rather than supporting it, they would have begun by looking for owls in previously harvested forests. They would have undoubtedly reached the conclusion that owls depend on certain structural arrangements of trees in forests, and that both old-growth and actively managed forests can provide these necessary structural conditions. They might have even discovered, as some have suggested, that some managed forests may be more effective than old-growth in producing owls.

However, a more critical approach to testing their original hypothesis would have required them to challenge a sacred assumption. They would have risked the guilt and shame of questioning the strict division between "forests" and "tree farms." Standards of purity, as well as the cynicism of political expediency in using the endangered owl as a "surrogate" for legally unprotected old-growth, have very effectively controlled the production of "scientific" information about the owl.

Even the liberal press is beginning to talk about how "the elite environmentalist class" is distorting science to exercise social control. Gregg Easterbrook concluded a recent article in the *New Republic* by calling for "honesty about owls." He stated:

> If it is eventually understood that affluent environmentalists with white-collar sinecure destroyed thousands of desirable skilled-labor jobs in order to satisfy an ideology and boost the returns on fund-raising drives, a long-lasting political backlash against environmentalism will set in. [68]

Constricting Language

Restrictions on the use of language are very effective tools for ensuring conformity. Totalist communication is often restricted to an abstract, all-inclusive jargon, in which misuse of words is a cause for making moral judgments about one's character. My failure to abide by the rules of restricted conversation exposed me to the forces of guilt and shame when I started talking about the needless suffering of loggers.

Appropriate use of terms can signify a person as part of the inclusive group, while inappropriate use of terms immediately locates that person as an outsider. Lifton refers to such jargon as "god terms" and "devil terms" because these simple words or phrases mirror the extreme emotional loading of the language. To environmental totalists, "ecosystem health," "ecosystem integrity," "biodiversity," "Ancient Forest," and other abstract and imprecise terms connote concern for the environment, and are used more to ensure control over thought and behavior than they are to denote important objects and relationships. In contrast, "loggers," "timber," "human needs," "tree farms," and "forest management," have become terms that connote a lack of concern for the environment, despite how specifically they might be defined.

One consequence of this emotional "loading" of language is the way it distorts and confuses scientific and professional communication. An open process of scientific exchange is impossible when technical language is prescribed and monitored with the objective of maintaining social control. "Ecosystem management" is a likely casualty of such emotional and ideological loading. It has become a "god term."

Return To Childhood

What surprised me most about shrinking and growing quiet was how comfortable I felt. I was helpless and dependent, just as I had been as a child. Lifton helped me understand why I felt this way—why I was tempted to exaggerate the omnipotence of those who wanted to silence me—why I secretly wanted to believe they were right. Seeking security by looking to others as wiser and more powerful is a habit we acquire as children and carry with us into adulthood. We long for omnipotent leaders (philosopher-kings) who will lead us through this period of cultural crisis and rapid historical change. Yet, each of us must ultimately assume the responsibilities of adulthood if we are to resolve the environmental problems that lie ahead.

Silenced By Fear Of Professional Annihilation

I stepped into full adulthood late last year when I refused to be transformed or silenced by the informal pressures brought to bear by environmental preservationists. I had started writing this book over two years ago, but lacked the courage to squarely face what I was writing about or even to say it in public. Guilt did not work on me, but shame did. I knew I would pay a big cost in my career by telling the truth as I saw it. Professional ostracism is a kind of social annihilation most will avoid at all costs.

My experiences with the Forest Ecosystem Management Assessment Team (FEMAT) helped me understand how moral persuasion transforms individuals into willing tools of centralized environmental preservationism. My need for freedom wouldn't allow me to remain silent any longer. I had to grow beyond my limited role as a social scientist and speak out as a citizen and a full person. Given the climate of environmental totalism, the weakness of my professional societies, and the politicization of science, the forest products industry was the only vehicle for telling the truth about FEMAT. I withdrew from FEMAT's planning process and shared what I knew about how it had distorted the truth.

Wearing My Father's Shoes

My ostracism from professional contacts followed immediately. The rumor circulated that "the industry had bought and paid for Bob Lee." The real message was never stated directly, but it was clear: keep your mouth shut and repent or we will annihilate you professionally. My response was to reclaim my freedom by writing this book. I now own my self and can stand proudly in my father's shoes. I am accountable only to God.

8

Haunted
Sanctuary

I believe in nothing. I have shut myself away from all the rocks and wisdom of the ages, and from the so-called great teachers of all time, and perhaps because of that isolation I am given to bizarre hospitalities. I shut the front door upon Christ and Einstein, and at the back door hold out a welcoming hand to little frogs and periwinkles.

Charles Fort
Apocalypse Culture

Many in my generation and the immediate generations to follow have lost faith in rationality, science, and traditional religion. They have grown cynical. Some are out-right nihilists, and deny that there is any objective basis for truth. This has opened a great void into which has rushed a diversity of strange creatures.

Intellectuals have invented an elaborate jargon to talk about this change. I will not join in the effort to paper over this void with fancy words, but will instead look at the creatures that have found the void a comfortable place to set up residence. I will also look at why these creatures have found this a suitable habitat.

I know many of these creatures on a first-name basis. They have *been visitors in my house as well. For a long time I smugly took pride in my freedom from the hidden forces I saw plaguing others. Then I began to see strange creatures lurking in the shadows. This taught me I had been fooling myself.*

Formal education and reading could not substitute for what I learned from personal experience. I had invited strange creatures to invade my house by leaving so much of myself undeveloped. I realized the void would be filled one way or another, and I had better start deciding how I wanted it to be filled.

How Education Kills The Spirit

When I searched back over my life to discover where I had started to follow a path that had left the spiritual side of me undernourished, I realized my formal education, especially higher education, had been terribly deficient. I found company for my assessment in a book by Page Smith, the noted historian and founding provost of the University of California at Santa Cruz: *Killing the Spirit: Higher Education in America.*

Smith laments the deficiencies of a higher educational agenda that intentionally denies the importance of religion to an understanding of societies and their history and to the full development of young people. Without abandoning his scholarly independence in exchange for religious doctrine, he lays the blame at the feet of what he calls "Secular Democratic Consciousness" that originated with the American Enlightenment early in the Republic:

> It was suspicious of authority, believed, for the most part, in unrestrained democracy and in majority rule. As opposed to the doctrine of original sin, it believed in the natural goodness of man once he was free from superstition/religion.[69]

Starting The Housecleaning

Smith walked unannounced into my house and showed me how to identify and clean out some of the creatures I had not seen. He reminded me of something I already knew but had chosen to deny. The social sciences, especially sociology, had from their beginning tried to replace religion, history, and certain aspects of the humanities with a human science. The founder of "sociology," Auguste Comte, even called sociology "the religion of humanity." Smith summarized sociology's mission as follows:

> The science of society, like any proper science, should develop the laws of social behavior that, if followed, would produce close to *ideal conditions of life for the faithful.* If enough information was collected on diverse cultures, *patterns must emerge that could be used in the reformation of society in general.* [my emphasis] [70]

Science applied to humans would "reveal" laws for "perfect living," so that human behavior could be efficiently engineered. An image of society as a regulated beehive was not far from my imagination when I first read sociology 25 years ago. But my house had too many empty rooms, and I failed to see the unwelcome visitors who I had allowed to enter.

Sociology's Jurisdiction Over The Soul

The most important thing I failed to see was the way in which scholars had divided the world into scientific and nonscientific realms. Rather than viewing science as one among many ways of learning about people and the world in which they lived, scientists had claimed control over reality itself—asserted that nonscientific ways of knowing were unreliable because they were based on superstitions, religion and irrational thought. This is called "scientism." Sociology comes into direct conflict with religion because it was especially ambitious, and claimed jurisdiction over the whole realm of human experiences outside of the physical and biological sciences. It claimed jurisdiction over the soul, and sought to explain all experience in terms of human social relationships.

Annihilation Of The Soul Creates A Vacuum

How was I, as a professional sociologist, to recover myself from the claims sociology had staked on my soul? How was I to remove unwanted creatures and put sociology in proper perspective as a valuable approach for understanding relationships between people? I had to learn to see how the void created by the annihilation of the soul can become a vacuum into which rush all kinds of hidden thoughts, agendas, and fanatical beliefs.

Smith was primarily concerned with the incompleteness of higher education, and the mistaken belief that science could replace religion in people's lives. He did not explore how people who free themselves from religion often fill the void once occupied by spiritual experience with dogmatic commitments to ideologies such as capitalism, communism, environmentalism, or science itself (scientism). He did not explain how those who reject religion often become fanatically religious about *their* claims to know "true reality." He underestimated the prevalence of academic absolutism, in which isolated experts claim full sovereignty and declare what is true and untrue, what people should think and not think, do and not do.

A Religion Of Nature Is Filling The Vacuum

I decided the best way to clean my house was to take the "scientific" study of human behavior at its word, and look at it as a religion. This would enable me to name the creatures in my house and bring them into the light where they could be seen and understood for what they were. I am finding many religions, but one is the primary focus of my study—a religion of nature.

To a large extent, this religion of nature has rushed into the void created by a successful attempt to blame the modern environmental crisis on Christian religion. I am repeatedly amazed how students in my classes repeat a litany condemning religion and touting science without realizing their argument is both *fundamentally* religious and non-scientific.

Blaming Genesis Again

University professors hungry to clear the "cobwebs of religion" from young minds adopted Lynn White, Jr.'s blaming of *Genesis* (See Chapter 1) and inserted it in the prayer book of an emerging religion of nature. This litany now occupies the realm of "sacred science." It is dogmatically repeated without thinking about it either as a falsifiable hypothesis or as a perverting influence on the spiritual development of young people who enter college with a strong foundation in religious traditions.

The "Sacred Science" Of Blaming Genesis

Yi-Fu Tuan, a noted geographer, challenged this hypothesis by examining how the environment was treated in China and other Asian cultures where Buddhism was the dominant religion. Critics of *Genesis* often point to Buddhism as an "environmentally friendly" religion. However, Tuan found that Buddhists were no more likely than Christians or Jews to protect their environment, and were often even more destructive.[71] Despite these and other challenges, the "sacred science" of the new religion of nature has protected Lynn White Jr.'s hypothesis from the hard reality of empirical evidence.

Young minds with gaping voids created by such "sacred science" do not see the mystical manipulation and loaded language that slips through the walls and takes up residence in their souls. I will explore the worship of forests as one manifestation of the new religion that is filling this void.

Worshipping Forests

Soon into my housecleaning I discovered a creature that was very hard to identify. I could not put a name to it and so could not see it. Then someone challenged me with a difficult question after I had finished giving a talk: "Is all this forest preservation a sign we are worshipping the creation rather than the creator? Is all this preservation stuff just a bunch of idolatry?" This set me to thinking. I was able to put a name to the mysterious creature: worship of forests.[72]

Four distinctive elements characterize the worship of forests: (1) flight from time, (2) sanctification of nature, (3) species equivalence, and (4) belief that humans are defilers and violators of nature. I drew these characterizations from a study of how German romanticism had evolved into a religion of nature during the 1930s: Robert Pois's *National Socialism and the Religion of Nature.*[73]

Flight From Time

Religion has always helped people cope with the terror they experience when they become conscious of how quickly time passes and how they might be harmed by uncertain events in the future. As humans, we are plagued with awareness of our own deaths and passage through stages of life. We also live in historical time—a shared experience of time punctuated by changes, threats, opportunities, losses, and triumphs. We find it disquieting to live in time because we worry about the future and regret past omissions and mistakes. Sometimes we are overwhelmed by guilt and anxiety. Other times we are emboldened by hope and visions of salvation. Not knowing what is going to happen, we fear the worst. Many of us are tyrannized by fear of an impending environmental apocalypse. In the words of Mircea Eliade, at times we are virtually paralyzed by the "terror of history."[74]

This terror drives us to seek a sense of immortality by participating in something larger and more enduring. In his insightful book, *The Broken Connection: On Death and the Continuity of Life*, Robert Jay Lifton notes four ways we seek a sense of immortality: biological continuity of a family, theological principles for guiding religious beliefs in life after death, creative works such as writing and the arts, and participation in the natural order of the world.

Lifton notes that these ways of participating in a "larger life" may permit people to escape the terror of history by entering a "dream time" in which past, present, and future are melded together in a timeless present. Flights from time have been especially common during major social dislocations or disasters, and following abrupt social and cultural change. He notes that mass movements and outbreaks of collective delusions such as found in the Middle Ages, modern China, and the youth movements of the 1960s are common when religious institutions fail or are ineffective at playing their traditional role. In all cases, perception of the world is altered and the ability to reason is inhibited.

Escaping Into "Ancient Forests"

The mystical imagery of the "Ancient Forest" provides an escape into a timeless present that also embodies an idealized past and a utopian future. The "Ancient Forest" is a symbol of immortality, endurance, and biological continuity that escapes the "terrors of history" by transporting its participants into a timeless realm of endless natural cycles. By embracing the "Ancient Forest," they can leave both time and their private selves and join in an eternal community of nature.

The old-growth forest exhibition traveling the country captures this meaning in its video with the refrain, "... immune to time, ecosystems continue to turn."

Having been terrified by the impending environmental apocalypse and made insecure by repudiation of inherited religious institutions, empty souls seek relief from fear and "green guilt" by filling themselves with the enduring "spirit of the forest." Mystical participation in the stream of never-ending biological cycles "cleanses" the guilt and "saves" the soul. Individuals literally "save themselves" by "saving endangered forests," and feel the terror of losing themselves—suffering annihilation—if the "Ancient Forest" is logged.

Sanctifying Nature

To escape from time into the "Ancient Forest," forests must occupy a spiritual realm that stands beyond any human questioning. They must occupy "sacred space" if they are to provide ways for participating in "sacred time." The re-enchantment of nature with spirit, or spirits, goes hand-in-hand with experience of a realm of absolute purity and goodness.

Secular society and modern economic institutions have developed because Christian traditions freed people from the belief that nature itself was sacred. By re-sacralizing forests and other objects, followers of the emerging religion of nature will invite other unanticipated creatures to join their households—creatures responsible for the intense personal and interpersonal conflict that has always erupted when sacred spaces are violated. Hence, re-enchantment of nature may be the hidden force releasing an emerging religious war over the environment.

Humans Are Just Another Species

One of the creatures I allowed in my house is the mistaken idea that humans are "just another species." This idea crept in unnoticed when I began studying ecosystems and treated people simply as biological organisms. A new radical form of ecological ethics called "deep ecology" has made this principle one of its sacred assumptions. The new ethics of "species equivalence" has been spreading from deep ecology to science, public policy, and even established religion.[75]

Species equivalence denies that humans have a capacity for spiritual development not found by other organisms. Hence, you can imagine how strange I found an Episcopalian clergywoman's statement that she believes deep ecology is "deeply inspired by the Holy Spirit." Had God suddenly chosen to speak with equal authority through all organisms? Were humans no more "spiritual" than frogs and periwinkles? What moral boundaries would protect us from the temptation to exchange humans for members of other species when our population has grown to a point where certain species cannot be preserved without "ridding ourselves" of "surplus" humans? Such dogma is already preached by some deep ecologists.

My broom worked even more furiously as I thought about how loggers and other wood products workers had been denied their rights to exist—had been morally excluded by scapegoating and shaming. Any idea that would lead people to trade off human dignity for sacred forests was named and expelled from my house.

Humans Are Despoilers Of The Natural Order

Species equivalence is connected to the nihilistic belief that humans are an ecologically corrupt species that has fouled its nest. Hence, humans are sometimes portrayed as a "plague" or "cancer" that is destroying the earth. Other forms of life are often thought to be superior to us because they are "natural" and do not destroy the environment. "Nature" must therefore be placed in reserves to keep it safe from humans—from ourselves.

This creature was easy to name: "the original sin of living." If humans are inherent despoilers of nature, then to live is a sin. At first I thought such ideas were restricted to deep ecologists and other adherents for the religion of nature. Then I heard Sallie McFague, a Protestant theologian, give a talk in which she extended Christ's Gospel Message to "endangered species, rain forests, and other threatened objects in nature"[76] by arguing that these were the new dispossessed and downtrodden and that it was sinful to harm them.

The "sin of living," especially when applied to the poor who have no choice but to subsist from land by cutting tropical rain forests, is a direct extension of species equivalence, and carries all of the same potential threats to human dignity and human rights. The guilt potential of this new ecoChristian theology (Or is it "Ecochristian" theology in the post-Christian world?) has environmental totalists lined up outside, waiting to squeeze through the walls in anticipation of a festive victory party.

Appeal Of A Religion Of Nature

Although I got to a point where I was ready to sweep my house clean, I fear many others have good reasons to get along with their new boarders. A religion of nature, especially when featuring "forests" as central icons, can do much to assuage "green guilt" and avoid the social pressure of potential ostracism. The worship of the "Ancient Forest" and "natural processes" could relieve people of the ecological terrors they have inherited from their predecessors. They could continue to effectively deny the ecological consequences of their own consumption activities by exporting ecological disruptions to other nations where such "dirty industries" as "logging" would be a welcome economic stimulus.

Denying Ecological Imperialism

By worshipping forest icons, Americans can relieve much of their burden of guilt and fear, and continue to consume energy, goods, and services at the highest per capita rates in the world. Ozone depletion, global warming, acid rain, pollution, and deforestation would remain hidden in the dark recesses of their minds. This shared illusion of purity would undoubtedly contribute to worsening international relations when the ecological imperialism, hypocrisy, and elitism of this religion became more evident.

"Ancient Forest" sanctuaries, no matter how necessary for ecological or scientific purposes, are likely to serve as the central focus of new "nature cults." Like most cults governed by totalism, forest worship would selectively interpret historical realities as signs affirming its own beliefs, rather than as historical events requiring rational assessment, scientific research, strategic actions, and collective sacrifices necessary to benefit future generations.

Taking Responsibility For My House

I continue to discover strange creatures in my house, and will undoubtedly find many more in the future. Some have refused to give up residence, but they respect the power of my broom and fear I will fully accept the higher power that sets it in motion. They know they will no longer be as welcome when that day comes.

I allow sociology to remain because it serves me well in understanding mass movements, group dynamics, and institutional structure. I also allow forestry to remain because it teaches me about the processes by which forests change and how we can work with these processes to meet our needs while protecting the environment. Sociology knows that it will be banished forever if it tries to reclaim my soul. And forestry knows I will sweep it out if it ever again invites its nasty cousin Holism to join it in my house.

9

Talking
To
Death

When I was a child, I spoke as a child,
I understood as a child, I thought as a child;
But when I when I became a man,
I put away childish things.

For now we see through a glass, darkly;
But then face to face.
Now I know in part;
But then shall I know even as also I am known.

1 Corinthians 13:11

They say the sixties' flower children never grew up. They just
grew wider in the hips and grayer on the head as they continued to
pursue their dreams. A childish aura of innocence—a playful attitude
toward reality—wraps many in this generation in a halo of unfulfilled
wishes for a better world. Yet, for many unrealized dreams of transformation have bred a hidden cynicism, and a readiness to control others
who stand in the way of utopian dreams. Moral persuasion provides a
tool for forcing transformation on others.

I could not help but be attracted to those who dream, especially *because their wishes remained unfulfilled. How could I be angry at children who played so freely with reality? I was caught and dragged along by the forward thrust of the currents they set in motion. I was trapped in whirlpools of childhood idealism and did not seriously think about being an adult until I was almost 50.*

No ceremony marked this passage. I just started feeling my age for the first time in 25 years and noticed I was not taking myself so seriously. I had somehow grown wiser and more accepting of myself. Unlike most previous cultures, our passages through the stages of life are not punctuated by rituals. Young men are not sent off alone into the wilderness, to return transformed by encounters with the spirit world and a new name conferring an adult identity. Although many Jews still practice public rites of passage, most Christians have abandoned community-wide rituals of confirmation in which childhood is vanquished and we are born anew as young adults.

I was stranded in my youthful idealism and did not know it—at least, not until I began to see that playing with reality had become a deadly game. I did not really grow up until I had talked to death.

While I never had much money, and had to save and work hard to achieve all I have accomplished, I had never suffered from hunger, terror, death of a family member, or faced imminent death myself. Encounters with threatening, unyielding realities often seem necessary to force people to let go of their childhood fantasies and face the responsibilities of adulthood. Because I lived a relatively protected life, I did not realize how common it has been for young people to become adults by talking to death.

There are many gates to adulthood. The realities of war and death in the immediate family are two of the most common passages. Surviving the horrors of war has for many men been the gateway between boyhood and adulthood. The vast majority never want to see war again, and many, like General Eisenhower, dedicate their lives to preventing it. Fear obliterates the innocence of childhood, and the horrors of war implant a mature respect for the hardness of reality. Soldiers leave home as boys and return as men. Those of us who do not go off to war, suffer similar horrors at home, or face dangers that bring us face to face with death, run the risk of remaining in our childhood dreams—

a world in which we cannot know reality because reality is still a plaything.

I was stranded in the cross-fire on a moral battlefield in a war declared by those who still see the world as a storybook contest between good and evil. I had thought it was just play—"just pretend," as I used to say when I was little. I was exhilarated by the rush of the wind against the banners and the shrill screaming of the charge. I was eight years old again, and we were leading the Crusades southward, marshaling our forces, and annihilating all heathen in our path. I carried the sword, swiping off heads on all sides, while three sisters cut a triumphal swath through the wet grass of the mountain meadow with a willow-pole crucifix and brilliant banners made from mother's worn-out sheets.

"This is war!" I shouted. "Yet this war is pretend, isn't it?" Doubts broke though the veil of fantasy and pressed relentlessly on the halo that had protected me throughout the battle. How could this be play when my sisters were nowhere to be seen, the mountain meadow was long forgotten, and I saw real tears and the dazed look of war-torn lives. Proud, hard-working people were milling about in confusion. "What is happening to us?" they pleaded. "What did we do to deserve this?" they asked. Many were without work and income. Some were hungry. Others were hungry and homeless. Families were disintegrating. Fortunes built on years of striving and saving were lost. Hopes were dashed. And fear was everywhere—a paralyzing fear that clung to the back of their throats and could not be coughed up, screamed out, or cried away.

My halo broke and fell to the ground. I saw reality though a glass, darkly. Death was stalking the battlefield and reaping fallen souls. Death rose from the mists and stared me in the face. I looked into the void. We talked for the first time.

Playing "Save The Earth!"

Death's first words released a cascade of questions, images, and raw emotions. What I thought was "just pretend" was "really real." Here I was in the middle of a battlefield, passively going along with a game we as children would have called "Save the Earth!"—a game I now call "moral persuasion." Then the dull grayness of death told me this was not play, but reality.

Do Loggers Symbolize Death?

But why play "Save the Earth!"? And why attack the "logger"? Helen Davis, the grand matron of Washington's southwest coast, and author of both Washington's State Song and a musical entitled "Eliza and the Lumberjack," accumulated two-thirds of a century's experience living among loggers and teaching their children to play music and sing. She once told me that loggers are "so close to death" that "nothing really matters to them." They are as "close to a blade of grass, a tree, a flower ... as they are to people." They participate in the stream of life: "They like to cook ... to make candy ... their perfume is the woods ... and yet, in way, they are already dead."

She was not simply referring to the hazards of their work which have blighted them with one of the highest injury and death rates of all occupations except mining. She was also referring to their inner lives. I took her to mean that they were too much a part of the forest—of nature—to fully join society and be accepted. Their work is so dangerous that they have frequently seen death in the woods when cables whipped uncontrollably, logs rolled, and equipment toppled, broke, or flew out of control. They have talked to death, perhaps daily, and death has taught them the hard realities of life. And for this reason they were at home with death, and like so many rural people accept death as a normal part of living.

Loggers were placed in a difficult position when others came to see loggers as agents of death because they "murder trees" or "destroy ecosystems." White urban residents living privileged lives have a difficult time accepting the reality that death is always and everywhere a companion to life. They are cut off from the life and death cycle of nature by elaborate technologies. The chance of violent death is so low for well-off college graduates living in safe neighborhoods that death seems unreal. Moreover, the threats of street crime and automobile accidents make death into a human-caused event, not a part of nature. This also leads urban residents to adopt romantic views of nature as "stable," "non-threatening," and "balanced." They have never had to survive a storm at sea, escape a raging wildfire, leave home by boat in a flood, or find their best sheep or cows slaughtered by predators. For many urban residents, people, not nature or chance, is the agent for death. This makes it easy for them to see the "logger" as a "killer."

Vanquishing Death By Killing The Logger

Loggers, portrayed as agents for death, became a powerful symbol in the war over the environment. They became a condensed symbol for the "frontier mentality" that had to be annihilated—a collection of symbols ranging from the "buffalo hunters" of the last century to those who are this very day "fragmenting ecosystems." In this fantasy world "loggers" were the unwitting agents of death—"tree murderers," "rapers of the land," and "destroyers of the earth." They symbolized the destructive part of ourselves that had to die for us to be "reborn" as members of a new "community of nature."

By playing "Save the Earth!" and attacking "loggers," many urban people felt they were vanquishing death itself, "saving" both the earth and their "reborn" selves for a life of joy, sharing, and playful intimacy. The innocence of well-off urban residents is reflected in a playful view that the hard realities of death could be overcome by eliminating those "bad people" who cause it. Annihilation of those who were seen as killing life itself would keep death at bay.

Building "Community" By Projecting Evil

But "loggers" were also more than this. They were also sacrificial objects upon whom were projected the shadows hidden deep in the "old self." They symbolized the dark sides people escaped from when they were "reborn" as enlightened participants in a "community of nature." And the ritual killing of the "logger" brought people together—a state of holistic unity—when they unwittingly expelled their inner poisons and denied it was their own shadow they saw in the "destructive" life of the "logger."[77] Participating in the symbolic murder of the logger made people feel whole, individually and collectively. The "logger's" death was their "re-birth" because it symbolized the death of their "old self" that had interfered with nature.

You Cannot Not Play

We were all included in the game of "Save the Earth!" regardless of whether we wanted to play. We could not escape. The media made sure of it by creating a totalistic, either/or drama of "good versus evil," of "owls versus loggers." A strangling double-bind silenced those who were urged to protest against such thought control. A refusal to play "Save the Earth!" was a sign that one wanted to kill owls, destroy wetlands, or foul the earth. We had to be either for or against saving the earth. Most of the national media did not give us the choice to be neither an "owl killer" nor a "logger killer." A reasoned voice was silenced and the public was not adequately informed that this game was not "just pretend"—real tears were shed and real lives were torn apart.

A Cleansing Ritual

But this was not all. Playing "Save the Earth!" made people feel they had cleansed the forest, and by saving the forest, had cleansed their souls. People became innocent again. They were washed clean. They were born anew. And the forest was clean and holy—sacred and filled with the life of spirit, now that death was being vanquished, now that the "logger" was dead—or at least transformed into a life-affirming agent for "restoring ecosystem integrity."

I watched the "logger" die in this "pretend" war. Although he may not yet know it, his life's blood has been drained, and his public life is over, annihilated by a game he never agreed to play. A symbolic murder has stolen his adulthood and drawn him into a "conversation" with the world of childhood fantasy. He is now asked to "play" at his work to avoid "destroying" nature. He is told he must repent for his former sins by "restoring" the ecosystems "destroyed" by his "frontier mentality." The abandonment to hard and dangerous work through which he talked to death has been replaced by the prattle of a "kinder and gentler forestry."

Refusing To Go To War

While the "logger" of the past is gone, or failing fast, those who log, ranch, fish, mine, and build things we need know too much about reality to escape into a world of childhood dreams. They know too much about life to pretend that death can be vanquished. They know, but don't talk much about how powerful leaders in the environmental movement are naively playing with death by forcing suffering on hundreds of thousands of rural people whose lives are torn apart by idealistic environmental reforms. They can clearly see that those who launched this assault on rural people have themselves become the agents for violence and death. And they know somewhere deep in their souls that death would triumph if they themselves become agents for death by using violence to answer the "war" into which they have been drawn. They have gained too much respect for the hidden power of rolling logs, falling limbs, whipping cables, storms at sea, collisions, disease, and natural predators to launch a suicidal counter-assault on these playful bearers of death.

Tell America's Story

But talking to death has also taught them that the agents of death will win out if they are not confronted. Although now too few in number to confront their aggressors by themselves, they can tell others their stories about the hazards of denying the reality of death. They can tell others about the lies these cynical sixties' flower children tell themselves and others. They can tell others about the game of moral persuasion and how it is played. And they can tell others about the suffering caused by childhood fantasies empowered by the forces of moral persuasion.

Rural producers can regain a moral stature that will make them tower above the aging cohort of increasingly cynical flower children and their followers. But they must first refuse to go to war with themselves or those who seek to transform them. Violent reactions to aggression and oppression will assure environmental preservationists that they were correct in marking them with death. To regain the moral high-ground, rural producers must take their story to the American people and demonstrate their commitment to finding peaceful resolutions to conflict and compromises through democratic governance. They must also regain the public's trust by demonstrating a commitment to responsible land stewardship.

The story rural producers can bring to the public is America's story. And because it is America's story it will triumph over the cynical manipulation of those who degrade rural Americans in an attempt to transform this nation into a collectivist state governed by philosopher-kings. The rural producer's story is about pride in honesty, hard work, self-reliance, community, land stewardship, and faith. The story is not perfect because real people are involved and mistakes have been and will continue to be made. But since this is an honest story, how could anyone be ashamed to tell it? Rural producers have always stood tall with courage and pride when they realized they were an essential part of the nation's conscience. That time has come again.

But my life story is also an American story of making choices. The war over the environment forced me to look at myself and to make a choice. I could continue to restrict thinking and writing to the scientific problems defined by the forestry and sociological professions in which I participated. This would bring me professional recognition and research funds. But this choice did not allow me to look at what mattered most.

My other option was to give life to the protest against the war over the environment that rose from deep in my soul. If I chose to protest this war I knew I could never go back because I would be forced to face myself and stop fighting the parts of my inner self I had long hidden from public view, and from myself. I would have to reclaim ownership of myself and take responsibility for developing the spiritual and intellectual gifts I had secretly carried all these years. I could continue to practice science, but would use science as only one among several methods for seeking truths.

I chose protest. Freedom followed, and I began the unfolding revealed in the above five chapters.

To continue with T.S. Eliot:

> Through the unknown, remembered gate
> When the last of the earth left to discover
> Is that which was the beginning;
> At the source of the longest river
> The voice of the hidden waterfall
> And the children in the apple-tree
> Not known, because not looked for

But heard, half-heard, in the stillness
Between two waves of the sea.
Quick now, here, now, always—
A condition of complete simplicity
(Costing not less than everything)
And all shall be well and
All manner of things shall be well
When the tongues of flame are in-folded
Into the crowned knot of fire
And the fire and the rose are one.[75]

Restoring America's Political Landscape

We feel ... moral strength because we know we are not helpless prisoners of history. We are free men. We shall remain free, never to be proven guilty of the one capital offense against freedom, a lack of staunch faith. ...

We must be willing, individually and as a Nation, to accept whatever sacrifices may be required of us. A people that values its privileges above its principles soon loses both.

These basic precepts are not lofty abstractions, far removed from the matters of daily living. They are the laws of spiritual strength that generate and define our material strength. ...Love of liberty means the guarding of every resource that makes freedom possible—from the sanctity of our families and the wealth of our soil to the genius of our scientists.

Dwight D. Eisenhower
First Inaugural Address
January 20, 1953

10

Getting
Real

'Tis surprising to see how rapidly a panic will sometimes run through a country. All nations and ages have been subject to them.... Yet panics, in some cases, have their uses; they produce as much good as hurt. Their duration is always short; the mind soon grows through them, and acquires firmer habit than before. But their peculiar habit is, that they are touchstones of sincerity and hypocrisy, and bring things and men to light, which might otherwise have lain forever undiscovered. In fact, they have the same effect on secret traitors which an imaginary apparition would have upon a private murderer. They sift out the hidden thoughts of man, and hold them up in public to the world.

Thomas Paine
The American Crisis

My protest against the war over the environment is rooted in conservatism wisdom. Like the American Revolution, in which colonists fought to restore the rights enjoyed by all free Englishmen, today's protest against the force of moral persuasion seeks to restore individual dignity and rights to the pursuit of happiness. I offer suggestions with the full knowledge that science is insufficient to provide answers to ques-

tions of freedom, and that this nation needs an open debate about the morality and politics of environmental management and protection.

The panic to "Save the Earth" has indeed produced as much good as hurt. The enormous suffering it has brought to rural people in the United States is not in vain. It has brought to light a crucial choice facing the American people by showing them that they are all participants in a new game. Whether they like it or not, all Americans are involved in a struggle with themselves induced by centralized attempts at moral persuasion. They can end this struggle in two ways: by exchanging their freedom for an idealistic promise of environmental security, or by rejecting the "war over the environment" and learning to take responsibility for caring for both other people and the environment. The strengths of American political institutions offer most of what is needed to exercise this responsibility.

This is not a choice between embracing a new faith or returning to a simpler time. Rural producers cannot go back to the "good old days." Compelling new realities demand our attention: world population is expected to double within 50 years, creating unprecedented needs for food, clothing, housing, transportation, education, and medical care; existing lands dedicated to production will be insufficient to meet the needs of this growing population; undisturbed (natural) lands will be exploited and/or cultivated and managed intensively to meet these needs; increasing numbers of species will go extinct and natural areas will be lost; pollution of air, water, and soil will increase and disrupt natural ecological functions and impair human health; new technologies will be developed to make resource production more efficient, increase utilization and recycling of materials, and reduce pollution and its harmful effects; and new information-based technologies will dramatically alter where we live, what we do for a living, and how we gain control over our lives.

Advocates for centralized control of the society and economy think these realities must be met by revolutionary transformation of both our citizens and institutions: curtailment or elimination of private property rights in land, forests, and water; elimination of profit-making from natural ecological systems upon which society depends; establishment of "ecosystem management" as a means for placing inter-

mingled private and public lands under the centralized control of the state; curtailment of pluralistic political debate in exchange for government-promoted communitarianism; reduction of local community autonomy to bring people into conformity with "ecosystem management"; and broadening of religious belief to focus ultimate concern on protecting the "community of nature." These transformations are being promoted as means for getting Americans to see that they must change from their old "exploitive frontier mentality" and come to see themselves as subordinate parts of ecosystems.

Advocates for central control are not evil, and most lack the stomach to employ the violence necessary to implement their utopian dreams. They are only effective because people have not discovered their game, or seriously considered the dangers of relying on moral persuasion to elicit conformity to their goals. Their danger is not just what they do to create a new society by messing with people's minds, but is also what they destroy in the existing society. By celebrating a biocentric "community of nature," they depreciate the value of human life and prepare the way for successors who may actually have the stomach for violence.

The choice is clear. We can chase our dreams and risk tyranny, or we can listen to accumulated wisdom which tells us that much of what we need to use our environment responsibly is already embodied in America's political traditions. Our traditions already contain most of what other people around the world who seek sustainability are trying to attain: individual initiative, self-reliance, respect for law and social order, commitment to strong families and local communities, and civil debate among competing interests. A weakening of these traditions is causing people to feel insecure and become vulnerable to moral persuasion, uncompromising positions, and violence. The answer is not to embrace revolutionary change, but to restore our traditional strengths by reclaiming ownership of ourselves and our institutions.

We have every reason to be hopeful. A growing number of people share a concern with managing our environment while protecting its most essential ecological functions. They are working toward what many call "sustainability" or "sustainable development." Sustainability is not a new idea. Europeans talked about it several hun-

dred years ago. The Greeks and Romans used other words to talk about it before the Europeans. And long before that the peoples of the Middle East told stories to talk about it. The Talmud recorded one these stories:

Chonyi the Magid once saw in his travels an old man plant-ing a carob tree, and he asked him when he thought the tree would bear fruit. "After seventy years," was the reply. "What?" said Chonyi. "Doest thou expect to live seventy years and enjoy the fruit of thy labor?" Said the old man: "I did not find the world desolate when I entered it, and as my fathers planted for me before I was born, so I plant for those that will come after me.

Millennia have passed and the basic message of sustainability remains essentially the same. We are not here simply to satisfy our own needs. Just as we inherited a world in which others worked to make life better for us, so we are obligated to protect and enhance our surround-ings to benefit future generations.

We use many different expressions to talk about this moral obligation.[79] Some say "We have not inherited the earth from our fore-fathers, we have borrowed it from our children." Others use the logic of modern capitalism and say "We must live off the interest (flows of resources) and protect the capital (basic resource stocks)." Some prefer fables and warn us "not to kill the goose that lays the golden eggs." In 1987 The World Commission for Resources and the Environment stated much the same thing as international policy when they said that sus-tainable development would "meet the needs of the present without compromising the ability of future generations to meet their own needs." All these expressions are talking about essentially the same thing.

All these expressions are also linked to the accumulation of wisdom on how best to motivate people to protect and enhance their sur-roundings for the benefit of future generations. This wisdom developed over the ages as people struggled to live with their surroundings. It offers us a precious opportunity to "get real" by building on the strengths of traditions that have been tested repeatedly by generations of practical experience. And as we shall soon discover, this inherited wisdom is consistent with some of our basic American traditions—espe-

cially families, communities, private property, free association, citizenship, independent science, and freedom of religion. It is also consistent with what social scientists have discovered about what makes possible the sustainability or conservation of natural resources.

Commitment To The Future

Let me begin with what has been learned about natural resources conservation. Walter Firey, a distinguished sociologist who studied natural resources, spent his career studying what makes conservation possible. His discoveries can be applied to sustainability, since both sustainability and conservation address the same basic problem: how people create a better future by taking actions in the present.

Firey referred to the basic issues that underlie sustainability when he stated, "there are many kinds of activities which ... require some kind of orientation on the part of human agents to a remote future."[80] He struggled with the same problems people have always struggled with when he said:

The cultivation of certain perennial tree crops, such as olive oil, cocoa, and pecan, presupposes many years of care before the cultivator will reap any marketable crop at all. Sustained yield management of forests in several European countries has involved reproduction cycles of more than a century. Amortization of capital investments in some mining and plantation enterprises often transcend the span of a single generation. Maintenance of soil fertility in peasant cultures, such as those of Europe and China, has imposed costs upon generations who have never realized any compensation for their trouble.[81]

Firey could also have talked about the great hydraulic civilizations of the Middle East and Asia whose irrigation projects required hundreds of years to build, but yielded relative prosperity for a millennium. There is vanity in contemporary discussions of sustainability which see it as a new idea and ignore the fact that the problem of creating a productive relationship with our surroundings is as old as civilization itself. After all, the development of settled agriculture required the building of irrigation systems and the building and maintenance of soil fertility.

Idealism Will Not Conserve Resources

One of the weaknesses of American society has been its tendency to substitute idealism for the wisdom of tradition. This is especially true of natural resource conservation. The ideal of natural resource conservation has long been a *moral imperative and has been unconnected to practical ways of motivating people to take better care of land.* In recent years the preservation of natural ecosystems has become a moral imperative engendering even greater idealism.

Yet what researchers have found most troubling is the gap between what people say and what they do. Although Americans share an idealistic commitment to conservation values, these ideals are not reflected in the behavior of people who manage natural resources or consume the products made from these resources. Studies of soil and water management show that conservation ideals are insufficient for motivating farmers to practice conservation.[81] Farmers seldom act on the idealistic conservation values they share with the nation as a whole.

Similarly, idealistic consumers—including many who identify themselves as "green consumers"—say they place great value on the future of their environment. Yet these same people are generally indistinguishable from the "greedy people" they disdain in their use of automobile transit, airline travel, wooden houses, suburban lots cleared from forests, and recreational uses of the outdoors. What all this says is that ideals are by themselves insufficient for motivating people to take better care of their surroundings. Additional conditions are necessary for well-meaning ideals of conservation or sustainability to affect every day behavior.

Three Things Needed For Conservation

Walter Firey's greatest contribution was to use science to demonstrate why traditional wisdom is still relevant to modern people. He showed that three conditions were necessary for people to conserve natural resources: resource management practices must be (1) *biologically possible* (ecological systems must be able to yield a flow of resources indefinitely—implying that ecological systems can bounce back in the face of natural and human-caused disturbances), (2) *socially and culturally acceptable* (people must *voluntarily* integrate conservation into the daily life of their families, communities, religious practices, and voluntary associations, and these values become part of their consciences), and (3) *economically gainful* (individuals, families, groups, communities, and other cooperating economic units must see some way of benefiting from voluntarily adopting the conservation actions).[83] Conservation is only possible when biological, social, and economic needs are met.

Firey used science to explain what had been common sense for a long time: to live with land over the long run, crops have to continue to grow, new ways of doing things have to fit with the way people live, and people have to get something out of doing things or they will not be motivated to work hard. But he added something very important to understanding conservation when he said that natural resource conservation also required people to be motivated to do things that are not gainful in the short run. He demonstrated they were most likely to sacrifice short-term gains for long-term benefits when their way of life and social institutions provided predictability and security.

Coercion Works But Is Expensive And Unstable

While heavy emphasis was placed on the need for conservation programs to build upon the traditions and values of local people, Firey also recognized that conservation programs had at times been achieved by relying on the authority of government to force people to adopt new resource practices. However, he was quick to point out that government coercion is both expensive and unstable. Others have explained why moral persuasion, as a more friendly form of coercion, would not work well.[84]

Enormous government resources are needed for maintaining surveillance of people using land, as well as for exercising the police force needed to keep people in line with government plans. Government coercion is unstable because people around the world are not easily controlled by moral persuasion; do not like to be ordered how to live; and tend to react to coercion by deviating from government policy, sabotaging government property, and even rebelling. These problems of social control further increase the need for surveillance, policing, and moral persuasion. A futile downward spiral of escalating coercion and manipulation is set in motion.

Local Control Is The Key

No society can effectively live sustainably with land unless it elicits a willingness of people to adopt new practices and hold one another accountable through informal, local social controls.[85] Conservation then becomes part of people's consciences and expectations for one another. International development work has been an important proving ground for learning how to motivate people to adopt conservation practices.

International Failures Of Top-Down Planning

Many of the failures in international forestry development efforts can be attributed to centralized (top-down) planning that imposed rational blueprints for radical change in land utilization and management.[86] Typically, centralized governments would ignore traditional customs and rights that local people had long had regarding land, trees, fruits, berries, and other resources. Government would nationalize ownership and employ local people in plantation forestry. The results have been uniformly unsuccessful throughout the world.

I was once visiting forests in northwest Thailand and repeatedly encountered government pine plantations that had burned as soon as the trees grew dense enough to carry a large fire. Local foresters denied the need for any assistance with fire suppression. I asked a local farmer why so many young forests had burned, and he gave me a convincing answer: "This has been our land for hundreds of years. We needed to graze our cows. The government's trees were in the way." Fire has been the weapon of choice for dispossessed rural people throughout the world. Even Israel, with all its military and police power, failed to control Palestinian sabotage of forests planted on contested lands.

I am currently working with a graduate student from India. He is studying a rural social movement in which Indian peasants are taking control of lands they had deforested to meet their immediate needs. These peasants are reclaiming contested lands the government nationalized and attempted to manage for commercial wood production to provide state revenue. Local people had grazed their animals and taken all they could get until the land was bare. Now that they have retaken control, they have organized committees, gained government recognition, adopted self-policing to reduce illicit grazing and tree-cutting, and are succeeding in bringing tree and grass cover back to the land.

Similar experiences throughout the world have shown that the replacement of top-down government controls by bottom-up planning and self-governance has led to greater success in conservation projects. People who learn how to work with local biological and physical conditions willingly adopt restraints on their uses of land, and do things that will provide future benefits for themselves, their family, or their neighbors. The results are by no means all that is desired, but far more has been achieved by empowering local people than by taking away their rights and then using police or moral persuasion to make them conform.

Institutional Stability And Voluntary Consent

These experiences in international development have reinforced both accumulated wisdom and the more theoretical discoveries of social scientists. History, experience, and scientific findings all point to the need for basic institutional conditions to remain stable, or to change at a rate slow enough for people to adjust without losing the ability to predict how they should act today in order to maintain an acceptable future way of life for themselves or those they care about. People in developing countries have not been willing to plant and tend long-maturing crops such as trees when the chances of realizing gains were diminished by unstable property rights, inadequate control over fire and grazing, potential disruptions to family and community life, and ideological movements that undermined their basic religious beliefs.

Rules Of Property Must Be Predictable

Erich Zimmerman, a noted resource economist, observed that,
All perennial culture, but particularly the planting of trees, rests on the *stability of social institutions*. No one would be foolish enough to spend a decade or more ... to build up an olive grove which can bear fruit for a century unless he feels rea-

sonably sure of a reward for himself and his descendants.[887] [my emphasis]

Property is one of the most important social institutions because it provides rules governing who has what rights to use, buy, sell, or dispose of land or things that grow on the land. Property rights may be held by private individuals, local communities, or a nation state acting for the people as a whole. Individuals may also hold rights to use communal or public lands, or useful objects found thereon. The most important thing to know about property rights is how predictable they are, not whether they are private or public. Threats to property rights, or uncertainty about their future security, has throughout history resulted in the abandonment of long-standing commitments to resource conservation. Examples include failures to maintain irrigation systems or to protect established trees from fire or illegal cutting.

Revolutionary Change Leads To "Cut and Run"

Commitments to institutions are weakened during times of war, revolution, and rapid social change. Some of the recent accelerated harvesting of small non-industrial private lands in the Pacific Northwest can be attributed to cynical short-range exploitation originating from a climate of increasing instability and uncertainty in private property rights. Well in advance of rising log prices, many proud owners with long-standing stewardship commitments quickly liquidated their timber assets. They were faced with the possibility that they would lose rights to future harvests if their lands were declared to be habitat for the endangered northern spotted owl.

Most importantly, there are no government plans to compensate land owners for financial losses resulting from possible restrictions of property rights from endangered species, wetland classification, ecosystem management, or other environmental concerns. This uncompensated transfer of property rights causes people to "get the timber while the getting is good." Like Thai farmers who set fire to the government's pine plantations in order to graze their cows on the fresh grass, institutional instability causes people to adopt short-sighted practices that work against conservation or sustainability.

Radical environmentalists, biological "scientists," and government functionaries have joined forces in a revolutionary leap of faith to impose environmental reforms. They place the stability of rural America and its environment at grave risk because they appear to be repeating the mistakes of state coercion and moral persuasion that have failed everywhere else in the world. The United States government and some "scientists" are so caught up in apocalyptic fears and romantic visions of living in harmony with nature that they have ignored the hard lessons learned in the rest of the world. This includes much of what our international experts have been teaching people in other nations who face similar rural environmental problems.

Maintaining Traditions Is Gateway To Future

The lessons of international development and history present clear pictures of reality. Walter Firey summarized these lessons in a succinct warning that we would be wise to heed:

Future-referring values...are tied to particular social orders whose eventual demise they are destined to share...Their possible re-institutionalization in another or succeeding social order must be wholly problematic.[88]

Some may find it expedient to transfer private property rights, revolutionize federal forest management, and create a climate of fear in which guilt and shame can be used to control people. But the resulting disruption of inherited institutions is likely to have the unintended effect of undermining commitments to the future. Moreover, rural people have proven to be especially resistant to moral persuasion, and are not easily coerced into conforming with the centralized desires of government agencies or elite groups.

Breaking the bonds of trust that make social order possible will end up breaking the land. The land will be left in worse condition than it would have been had social order been maintained. How tragic that radical reformers could inadvertently accelerate the ecological disruption they seek to avoid by failing to appreciate the strengths of their culture and its potential for successful accommodation to new environmental realities. Much could be learned by reviewing the history of countries like Nepal, where disruption of traditional communal rights led to social breakdown and unrestrained exploitation of trees, grazing lands, and firewood.

Guiding Principles for Sustainability

The promise in living sustainably with our surroundings is to be found in the strengths of our traditional institutions, not in revolutionary transformation. The next four chapters will summarize seven basic principles that can guide us:

(1) **Secure property rights are essential for people to make sacrifices for the future;**

(2) **Voluntary conformity with conservation programs is possible if local citizens are involved in developing and implementing programs and are allowed to capture economic gains sufficient to maintain their way of life;**

(3) **Informal social controls embodied in strong families and viable local communities are far more effective than reliance on centralized police powers or moral persuasion;**

(4) **Local populations are vehicles for, not obstacles to, attaining conservation goals;**

(5) **Citizens empowered with knowledge, responsibility, and authority can make them reliable, effective, and humane agents for carrying out environmental reforms;**

(6) **Science as a method for maintaining a critical view of all factual claims can discover unforeseen choices and anticipate unintended consequences; and**

(7) **Freedom of religion can maintain the independence of government and remove it from its current role of promoting the spirituality of land, forests, or nature.**

I will summarize these principles by discussing why our traditional institutional strengths are so essential to solving environmental problems. Discussion will begin with family and community, and then turn to issues of political liberty, science, and religion.

11

Celebrating
Sally's
Heart

Advocates for centralized control are using the same approach to local communities as they use on individuals. They assume the role of "teachers" and seek to transform the local way of life in small towns from "frontier settlements" that depend on unstable extraction industries into "environmentally friendly communities." This was revealed in the way some of the "scientists" participating in the Forest Ecosystem Management Assessment Team (FEMAT) talked about "timber junkies" when discussing "solutions" for "timber-dependent communities." Government agents were there to help solve a problem of "addiction to timber." The rules of moral persuasion are clearly recognizable in the approach the federal government is taking to rural communities.

— 149 —

Internal and external struggle is created by undermining the "old way of life" and idealizing a "new way of life" in which there will be "harmony" between people, as well as between people and nature. Some local residents seek the advantages of aligning themselves with the new ways of doing things and get into conflicts with those who stick to the old. Federal and state agencies approach community leaders with "new programs" designed to encourage their "realignment" with the new social and economic vision. Banks and investors shy away from making commitments in these communities because the old rules of investment no longer hold when federal resource supplies are suddenly withdrawn, environmental regulations are imposed, and property rights are curtailed. Social service and research entrepreneurs emerge and seek new clients created by the decline in the local economy.

All this contributes to further conflict between local citizens. But, most tragically, it also undermines the security and predictability of local institutions. The social fabric begins to unravel, and the traditional strengths of small town life are severely weakened. At this point, many local citizens are sufficiently demoralized that they are ready to assume a new identity, and, sure enough, the government is there to help with the transformation!

World-wide experience suggests that maintenance and strengthening of the local social fabric is the most important thing that can be done to promote sustainable living. A highly developed sense of community is the foundation for sustainability. And families are the foundations for community sentiment. Families are the building blocks for community sentiment because individual character is formed by families that teach the values of self-respect, respect for others, self-discipline, deferred-gratification, and careful stewardship of material things. Families are needed to form a social and environmental conscience, and a sense of individual responsibility.

Community ties are the building blocks for sustainability because informal relationships set the expectations for how people will take responsibility for one another and their surroundings. Land stewardship is most highly developed when people share an ecological conscience and hold one another accountable for caring for one another and their environment. That is why I will tell you about Sally's heart.

Sally Sauer is a young woman who lives in Libby, Montana. Libby (together with the nearby village of Troy) is a remote settlement in northwest Montana which has long depended on logging, sawmilling, and mining. The wood products industry was developed with the encouragement of the federal government, which was committed to allowing use of national forest lands in the surrounding territory. But things changed. Job and business losses in the wood products industry have increased sharply in the last five years as a result of federal laws that require protection of grizzly bears, biological diversity, and environmental amenities.

Loggers and other wood products workers also suffered from verbal abuse at the hands of the press and radical environmentalists. Confusion, frustration, anger, hardship, and suffering of displaced woods workers had split the local population into warring camps. Trust had declined and hostility has been building beneath a surface of civility.

A couple of years into this crisis, Sally (whose father had worked for a local sawmill) was infected by a virus that weakened her heart muscles to the point where she was not expected to recover. A heart transplant was recommended as the only way to save her life. But, tragically, she lacked insurance coverage to pay for such an expensive operation and began to face the reality that she might soon die. Family members began to make plans to sell their homes to pay for the operation.

Some might continue the story by reporting, "This is where the community came to the rescue...." But this would mislead people into believing that "the community" was a unified thing that could act in a heroic way. That was not at all true. What really happened was that caring individuals living in Libby learned about Sally's tragedy and began to search for ways to help pay for the operation. Managers in the local Champion International sawmill learned of Sally's need, and offered to pay local logging contractors $ 50,000 to cut a special block of timber from company lands so that the contractors might donate these earnings to Sally's heart transplant fund.

A "sense of community"—built on shared concerns and the spirit of cooperation—emerged as others began to volunteer their labor,

resources, and equipment. Loggers jumped on the bandwagon and vol-
unteered to cut and haul the timber. Millworkers volunteered to work
extra shifts and saw the logs into boards. Champion managers arranged
for the mill to run additional shifts. Local businesses contributed and
helped organize other fund-raising activities. Service groups and
churches pitched in. There were bake sales and community raffles.
Local middle school children welcomed the first loads of log trucks as
they rolled into town by lining up along the main street and wearing spe-
cially-made paper hearts as a symbol of love for Sally Sauer.

People held their breath and prayed as Sally struggled to stay
alive while she waited for a new heart to be donated. The call announc-
ing a new heart came just in time; the transplant was a success, and
Sally returned to Libby to recover and take the front seat as the grand
marshal in Libby's Centennial celebration parade and serve as the
emcee at the Logger Days Queen Coronation. The spirit of community
caught hold. A permanent emergency medical fund was established in
Sally's name, and two more children have received heart operations
with help of this fund.

Freeing The Spirit Of Community

Sally Sauer was supported by a spirit of community cooperation
found in thousands of settlements throughout America. She is unique,
but her story symbolizes strengths that have built rural America and
can carry us through troubled times as we learn to take greater responsibility for managing the environment.

Caring people in Libby and Troy helped heal their wounds and
ease their pains by working to save Sally's life. Commitments to work
together to improve the management of our environment can also help
heal the wounds inflicted on rural loggers, sawmill workers, ranchers,
farmers, miners and others who have suffered verbal abuse, unemployment, and economic losses. The spirit of cooperation is still there, but it
must be freed to do its work.

Some rural people have been hurt so deeply that they call for a
"counter-revolution" to free people from an increasingly oppressive
government. I do not think that is a sensible answer. Revolution implies
radical change in institutions, and that is just the opposite of what is

needed. The needed institutional structure is intact. We simply have to learn how make it work for environmental protection as well as for resource production.

We could best begin by listening to the people most directly affected. While their answer appears simplistic—"less government"— there is a hidden wisdom in giving rural people both the authority and the responsibility to do a better job of managing the environment. Unlike experiences in other nations, there has been little experience in the United States with delegating authority and responsibility for environmental protection and management to local jurisdictions. There has been a lot of local authority for deciding on how to use land, but responsibilities for environmental protection have not been adequately coupled with these authorities. And where responsibility for protecting the environment has been attached to localities adjacent to federal lands, it has been coupled with moral persuasion and has, as a result, been highly unstable.

Moral Persuasion And National Forests
The history of the U.S. Forest Service shows how government worked with rural people to elicit commitment to conservation and local cooperation. No agency of the federal government, with the exception of the military, has been as successful as the U.S. Forest Service in eliciting the voluntary conformity of citizens. The Forest Service faced what appeared to be an impossible situation when it took over management of reserved federal lands early in this century. Uncontrolled grazing, mining, timbering, and fire were threatening the nation's new national forests.

Forest Service Chief Gifford Pinchot and his foresters exercised strong leadership, and did not hesitate to use law enforcement against violators. But the primary mission of the new agency was to work with rural people and elicit their commitment to conservation programs. Large mining, grazing, and timber interests stood to lose from redistribution of rights to local people, down-stream agriculturists, and the public as a whole. Pinchot was successful in defeating these powerful Western interests because he fought for the "little man"—the "home-builder" (small-town settler), families, and others who were seeking to find a better life for themselves in the rural West. These settlers received rights to use federal lands in exchange for cooperating with the government in protecting its lands from fire, poaching, and illicit timber cutting and grazing.

An Inspirational Example For The World

Within a generation, the Forest Service had eliminated most of the lawless activities and elicited the active support of local people. Since local people benefited from using the national forests, they tended to monitor one another's activities and reported troublemakers to district rangers. Few rangers were needed because they generally enjoyed the support and cooperation of enough local citizens to secure the establishment of conservation programs on tens of millions of acres spread throughout the West.

Successful implementation of federal conservation programs over such a vast area with only limited staff was an unprecedented achievement. It was only possible because the Forest Service had magnified its effectiveness by granting access and use in exchange for local support. Informal social controls arising from the locally-initiated voluntary cooperation of local people made the U.S. Forest Service a world-renowned success in government administration.

Excesses Of Timber Production

But things began to change in the 1950s. Increased emphasis on timber production accompanying the post-World War II housing boom resulted in the "industrialization of the forestry profession." Economic considerations crowded out long-standing commitments to maintaining social acceptability, as well as weakening commitments to maintaining biological stability. Congress and the Administration's Office of Management and Budget increasingly emphasized revenue production from federal lands. The Forest Service abandoned its tradition of inspirational leadership and began sending its chiefs off to the Harvard Business School for administrative training.

By the 1960s and '70s, extensive road systems had been built and extensive clearcutting of timber was clearly visible to the general public. Such intensive timber management activity appeared to contradict the intent of multiple resource planning embodied in the Multiple Use Act of 1964 and, later, the National Forest Management Act of 1976. Building on the generalized reaction to the anti-environmental rhetoric of the Reagan administration, many citizens concluded the Forest Service "had been captured by the industry." The national forests became center-stage issues in a national debate about the environment which continues to rage.

Paralysis By Procedural Revolution

By the mid-1980s the Forest Service was under an all-out attack by both radical and moderate environmental advocacy groups who were concerned more about prohibiting all timber harvesting than in making timber programs stay within the limits of sustainability. An accumulation of lawsuits brought federal timber sales to a virtual standstill on the West Coast in 1993, and reduced them substantially in the rest of the nation. Court injunctions issued to force the Forest Service to follow federal law in protecting the northern spotted owl were by far the most publicized actions.

Over the last 35 years the Forest Service slipped from the pinnacle of public respect to a state of confusion and demoralization usually only found in a routed army. Trust earned over a half-century was suddenly broken, and the Forest Service came under the close scrutiny and control of the Congress and environmental advocacy groups. The agency is currently attempting to "reinvent" itself and find a new center about which to rejuvenate morale and refocus programs.

Betrayal Of Pinchot's "Home-Builders"

Local residents who depended on federal timber sales to support jobs, businesses, and incomes felt betrayed and abandoned by their government when timber sales were so suddenly halted. After all, the government had initiated the timber sales program, assured the public that timber harvesting was regulated by long-term sustained-yield plans, and promised a continuous supply of wood for the indefinite future. Only in the last few years has it become public knowledge that timber harvesting on many national forests had been in excess of what could be sustained over the long-run.

As it had from its beginnings, the government had elicited support by encouraging people to locate in rural settlements, put down their roots, build their homes, invest in businesses, and count on continuity in wood supplies. Banks loaned money to local entrepreneurs on the basis of the government assurances. Thousands of dedicated and hard-working citizens found themselves "out on a limb" without any help when the timber sales program was enjoined. Local workers soon found themselves out of work, and business people went broke and lost personal fortunes accumulated over years of hard work. Moral persuasion, although relatively mild by world standards, had proved to be a relatively unstable means for assuring local social and economic stability and maintaining environmental quality. How stability can be re-established is now a pressing question. Especially questionable is the stability of voluntary social control in rural communities where moral persuasion is maintained and rights to use public lands are withdrawn.

Clinton Sets New Directions

President Clinton was exposed to the intensity of human suffering during his 1992 campaign, and promised to hold a "Timber Summit" to resolve the issue if he were elected. As one of two sociologists invited to participate in the president's "forest conference," I have had the opportunity to observe first-hand the unfolding of a new federal government policy for working with rural people who have relied on federal forest lands for their livelihood.

The Forest Ecosystem Management Assessment Team (FEMAT) plan recommended to the President would reduce timber harvesting on federal lands by over 80 percent, and contains special provisions for working with local people impacted by harvest reductions. Major emphasis is placed on the biological goal of restoring ecosystems to their "pre-settlement conditions" (biological conditions that existed when European settlers first arrived in the West). Re-employment of dislocated wood products workers in ecosystem restoration projects is a primary recommendation.

Restore A Pre-Settlement Community Of Nature

The Forest Service has entered a new era in which its leadership is readjusting its goals to a biocentric vision of restoring the forests to what they were in pre-settlement times. Like Pinchot, the new leadership faces the daunting challenge of a massive redistribution of rights. Only this time the plans are to shift rights from humans to other species. Yet what this shift in rights really does is to transfer uses and benefits of forests from long-standing users to citizens who prefer that forests be restored to presettlement conditions. This transfer can cause social instability when there is inadequate evidence to conclude that total restoration of forests is necessary for protecting endangered species. Pinchot's vision of productive human settlements existing permanently with their surroundings has been replaced by a vision of a "community of nature" in which humans are simply a subordinate part.

Using Moral Authority Of Ecological Science

Yet, unlike Pinchot's challenge, the new leadership will find it especially difficult to elicit voluntary conformity of local citizens to this massive transfer in rights. Personal dignity, jobs, businesses, and assets totaling in the tens of billions of dollars have been taken away from people whose proud way of life had been established by implementing Pinchot's vision. Trust had been eroding for 25 years, but was broken when the federal government suddenly abandoned its long-standing commitment to local people. The government has alienated those it had made dependent upon it, and must now draw on new sources of support to legitimate and implement its restoration plans.

Biological "scientists" have stepped forward and claimed the moral authority for managing the "community of nature." They claim their knowledge, and "humility" in the face of complex ecosystems qualifies them to "speak for the whole." They claim to "understand the problems" of local people whose lives have been disrupted by termination of federal harvesting, and promise to provide replacement jobs in forest restoration projects. And, like Stanley Milgram's subjects, the American people appear to have deferred to ecological authority and agreed to go along with restoration plans.

Local People As Partners In Experiments

The government's vision for reinventing rural society revolves around "Adaptive Management Areas" (AMAs) recommended in the President's plan for Northwest forests. These special administrative units would provide federal scientists with the opportunity to experiment with "ecosystem management" across both public and private lands, and would employ local people in ecosystem restoration or experimental work. AMAs would also serve as experiments for creating "partnerships" between local citizens and government scientists and ecosystems managers.

These reforms are guided by an idealistic vision of consensus groups that would be established to elicit the voluntary cooperation of local citizens in restoring damaged watersheds and helping implement ecosystem management. Partnerships between government ecosystem managers and local people are proposed to take the place of the gridlock originating in debates between the timber industry and environmental interest groups. Contentious debate characteristic of an open, pluralistic political system is seen as a threat to stable ecosystem management.

Reinventing Rural Communities

Social scientists are assisting biologists in "reinventing" forest-based "communities." Their contributions are inspired by what sociologists refer to as the "communitarian ideal." The communitarian ideal has generally been welcomed by both liberals and conservatives as a healthy reaction to the excessive individualism of American culture. Emphasis is placed on "communal relationships" and the "associated responsibilities they bring," as well as "virtue" and other ideals to be realized by communitarians. While I believe the communitarian movement has been important in reminding Americans of their responsibilities to other people, far more thought needs to be given to how it will be implemented—especially when the federal government takes an interest in promoting it.

I was exposed to the communitarian movement while studying and working at the University of California at Berkeley, and have followed its influences on the study of rural communities. A Berkeley sociologist who was involved in developing the President's plan had in earlier work justified communitarianism as a model for forest communities. He said:

An individual should not and can not stand apart from her or his community, particularly if some of the most important values flow from it. The community and its traditions must therefore be seen as informing the evaluation of [individual] well-being.[89]

Dangers Of The Communitarian Ideal

The belief that the "community" is a thing in which the individual is simply a part dovetails perfectly with the biologist's "community of nature" in which humans are simply subordinate parts. I have come to see the communitarian vision of "community" as an idol whose "power" justifies an expansion of government control over individual behavior, and I am not alone in making this interpretation. Even leading communitarians are aware of the dangers inherent in subordinating individuality to the whole. Christopher Lash acknowledged such dangers when he said,

"... the communitarian tradition, even though it appeals so powerfully to the sociable impulses destined to be frustrated in a competitive, individualistic society, remains suspect. From the time of Plato onward, its social ideal contains unmistakable authoritarian implications. Experience indicates, moreover, that the republic of virtue in practice issues in a reign of terror.[90]

The monopoly of moral authority claimed by biological "scientists" who seek to engineer local communities may simply be a more sophisticated version of moral persuasion. And the arrogance that leads some people to dictate to others how they should live is a long way from the "community spirit" that replaced Sally's heart.

Local Autonomy Can Be Restored

I believe it is not too late for the American people to do their own "adaptive management" by restoring local political institutions as the foundation for managing the environment. If the strengths of the American experience are not sufficient to convince people this is what needs to be done, then we can draw on what has been learned from reputable social scientists, international development projects, and the successes and failures of various governments and civilizations.

As the fall of the Soviet Union has demonstrated, even the most coercive top-down, command and control approaches will not work unless citizens are empowered and have a significant amount of control over their own lives. Europe, in particular, offers numerous examples of how the autonomy and empowerment of local people is the key to sustainable natural resources management.

The next chapter will explore how such decentralized social control can be achieved through altering the federal land tenure system. I will close this discussion with suggestions for how the local social fabric can be restored and strengthened in preparation for such institutional reforms. This is an important step toward land tenure reforms because strong families and communities are the foundations for sustainable natural resource management.

Restoring Local Institutions

Concerned individuals can help restore local institutions to their proper place in making responsible environmental management a central commitment of American life. As citizens we can:

(1) **Use local churches, synagogues, temples, neighborhoods, social service groups, and other local associations to organize forums for bringing people together in face-to-face discussions of issues surrounding environmental stewardship.**

Discussion topics might include (a) the ethics of making trade-offs between humans an other species, (b) responsibility of urban consumers for causing impacts on rural settlements and occupations adversely impacted by environmental reforms, (c) hidden interdependence of consumers and producers in a complex economy, and (d) how to become more responsible consumers and producers through buying less, recycling, using things longer, and honoring responsible producers.

(2) Organize professionally facilitated face-to-face mediation sessions involving radical environmentalists, moderate environmentalists, farmers, wood products industries workers, ranchers, developers, and others who use lands to serve human needs.

The purpose of these discussions would be to break down stereotypes and encourage people to humanize one another. Facilitators will be more successful if they understand how Calvinist and Social Darwinist beliefs have led people to take sides in warring camps and project evil on the "other."

(3) Organize professionally facilitated discussions among government employees working in land management agencies, government and university scientists, and local residents in the communities affected by government decisions.

The purpose of these meetings would be to restore and sustain the active participation of government employees and scientists in local social life—to encourage them to avoid the hazards of becoming outside social engineers or police agents, and to delegate authority and responsibility and strengthen local institutions.

(4) Build, join, or strengthen family support networks, including Big Sisters and Big Brothers, to assure that children receive the moral guidance they need to become responsible citizens.

The problems young people face are many, and are not limited to drugs, alcohol, crime, and sexual misconduct. They benefit from special help in learning how to make sound ethical judgments so that they will not be easily swayed by propaganda from any radical environmentalists or their counterparts (who deny the reality of all environmental problems).

(5) Strengthen our families by making "quality time" for spouses children, grandchildren, and others.

Families build strong moral character and the capacity for

reasoned ethical judgments. Responsible consumption and production can be best taught by living it. Loving families make
loving people, loving people will build community spirit, and
community spirit will maintain a healthy environment.

**(6) Celebrate the spirit that helped replace Sally's heart by
reaching out to neighbors in time of need, standing up for
what is right, taking pride in honesty and hard work,
becoming a responsible consumer and producer, and
demanding social justice for all people.**

All of these activities can help restore trust among people
who have been separated by fear, confusion, and misunder-
standing. But none of them will succeed unless government is
also restored to its proper place in American life, and is removed
from the business of controlling people's minds. I will now turn
to the challenges of "bringing government back home."

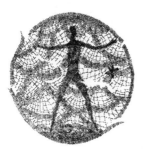

12

Bringing Government Back Home

Society is indeed a contract...partnership...the ends of such a partnership cannot be obtained in many generations, it becomes a partnership not only between those who are living, but between those who are living, those who are dead, and those who are yet to be born.

Edmund Burke
Reflections on the
Revolution in France

Research and international experience suggest that the best way to encourage people to live sustainably with land is to give them more control over their lives. Extending or restoring land or resource rights to local people is fundamental to giving them more control. Land and resource tenure provides predictability and has the advantage of coupling authority and responsibility at the local level. Yet leaders of environmental advocacy groups strongly oppose any attempts to extend rights to local people.

Why are leaders of the environmental movement so afraid to invest rural working people with the authority and responsibility to steward the natural environment when this is exactly what they would *recommend for rural people in Asia, Africa, or Latin America? Calvinist splitting of humanity into the pure and the impure and an ideological bias against capitalist forms of economic organization, together with their need for an enemy, undoubtedly has something to do with this contradiction. But equally, if not more important, is the control they would give up if they share any of their power with rural people.*

The environmental movement is in the driver's seat of a massive, self-propelling conflict industry that both raises money and extends social control. Environmental executives sit behind big desks in big offices, make big money, ride in big cars, sit at the table with big people, enjoy the blessings of big "scientists," rent votes from big politicians, and hire big staffs to invent big threats that will bring in more big money. Cooperation, compromise, or the sharing of power would weaken their control over the apocalyptic engine that keeps them going. For many leaders, improving the environment, to say nothing of "saving" it, becomes secondary to attracting more money and buying more influence. This is a major reason for political gridlock; it is not in their interest to give an inch, and their zealotry blinds them to the humanity and creativity of the local citizens who stand ready to practice environmental stewardship.

Sociologists have studied these problems in institutions as diverse as businesses, universities, and churches, and are familiar with how the goals of building and maintaining an organization take the place of the organization's original mission. In an open society, where competing groups can seize issues, excessive emphasis on maintaining power can make an organization vulnerable to losing control of the issue it was formed to address. That moment has arrived for the environmental conflict industry.

Rural Americans can bring government back home by seizing the issues of environmental quality and sustainable development and putting these principles to work in their lives and on their lands. The federal government will soon learn it does not need big environmental advocacy groups to get the job done. Millions of little environmentalists working in every village, farm, field, and forest can do a far better job for a lot less money. But what rural America needs to reduce top-down environmental regulations are inspiring demonstrations of its steward-

ship potential and workable ideas for assuring that this potential will be realized. Inspiring demonstrations and ideas are emerging daily, and are only in need of publicizing.

Take "Doc" and Connie Hatfield of Brothers, Oregon as an example. These ranchers have found a away to improve the quality of the range, protect soil and riparian zones, and make a good profit from cattle grazing by selling beef in specialized niche markets. They adopted "holistic resource management" practices developed by Allan Savory of the Center for Holistic Resource Management in Albuquerque, New Mexico. Cattle are used as a tool for stimulating plant growth and thereby restoring biological diversity and a protective cover of vegetation to prevent erosion. Cattle are carefully managed to keep them from staying in the same place for long periods of time. This practice mimics the nomadic grazing habits of ungulates which occupied the grass lands of the West prior to European occupancy.

The Hatfields are showing that it is not necessary to remove cattle from Western range lands to protect or restore the environment. Advocates for removing cattle grazing from all public lands have threatened a way of life followed by ranchers for several generations. New grazing practices can protect and restore biological diversity, protect riparian zones, perpetuate a valuable way of life, and allow ranchers to stay in business.

Private citizens working on private lands are already doing as much, or more, to promote sustainable management of the environment than government functionaries, biological "scientists," and environmental activists. These citizens are successful because they are building on traditional American strengths that have passed the tests of past crises and been modified to fit new circumstances. Two institutional strengths—property and practical knowledge—deserve special attention because they offer distinct advantages for solving environmental problems.

Private Property Is A Strength

The right to own private property remains one of the most important strengths of the American tradition. Yet property ownership is under unprecedented assault by advocates for environmental reform.

The list of restrictions being placed on private property rights seems endless: endangered species habitat classifications, wetlands regulations, water quality regulations, and "ecosystem management" requirements are only a few examples.

The curtailment of private property rights threatens to turn one of our major strengths into a weakness. Like small non-industrial forest land owners, people who are faced with losing their property rights may grow cynical and "get what they can get while the getting is good." Too much uncertainty about property ownership can cause people to liquidate their assets in ways that cause substantial damage to land and disrupt ecological processes. Owners who are in a liquidation mood are far less likely to practice careful stewardship.

Government Ownership As A Cause For Suffering

Environmental disruption and social and economic suffering also results from uncertainty about rights to use public lands. Advocates for revolutionary change blame private property owners, technological change, and "the industry" for creating economic chaos. Past mill closures, timber liquidations, and gains in labor productivity, no matter how disruptive to a local area, were relatively small disturbances compared to the massive social and economic disruption being created by federal government decisions to suddenly terminate timber harvesting over a vast three-state region—if not all federal lands. The conclusion is clear for all to see: government ownership, not private ownership, has been the most destabilizing influence on social and economic life in the Northwest.

The total cumulative impact of lawsuits, court injunctions, and ecosystem management plans will be enormous. By withholding so much wood from the market, tens of billions of dollars in timber asset values were transferred from the federal government (and therefore from the U.S. treasury and the U.S wood products job market) to other domestic and foreign owners.[91] Politically-induced shortages in softwood timber supply have caused prices to rise to unprecedented levels. As a result, private timberland owners, especially the large land-owning companies, are enjoying unearned wind-fall profits. Even foreign suppliers are capturing the gains. A representative of a New Zealand timber company recently asked a Northwest counterpart for a stuffed spotted owl so he could place this "icon" in his company office where all could "worship it."

By the end of this shake-out as many as 200,000 workers supported directly or indirectly by government-dependent timber enterprises are likely to have lost their jobs.[92] Although many have found other work, or will soon do so, the majority will suffer from reduced wages, unemployment, and early retirement on a reduced pension. And often overlooked in this debate are the family members of displaced workers, numbering more than 600,000, who will have their lives disrupted. Additional impacts will result from the failure of local businesses which are directly or indirectly dependent on government timber enterprises. Families, the strength of our future, and the basis for forming citizens capable of sustainable living, are being devastated by political and legal decisions to suddenly terminate timber harvesting on federal lands.

Delayed Impacts More Severe

The impacts expected to result from termination of federal timber sales have not occurred as suddenly as first expected. Instead, massive job, income, and business losses have been spread out over several years by rapid harvesting on small and large private ownerships. The decisions of small non-industrial forest land owners to liquidate timber assets before restrictions are imposed and while prices are high will result in a delayed impact of unprecedented proportions. And there will be no cushion when this panic of harvesting is over. John Beuter, a forest economist and expert on the Northwest's timber supply situation, is fond of saying that "we are eating our seed corn" and will suffer from many lean years until timber stocks recover. All this instability was triggered by well-intentioned people who set out to "revolutionize" forest management in the Pacific Northwest.

Politically Unstable "Elite Settlement"

Instability in rights to private and government property creates a hazardous political situation. Large land-owning industries and the environmental community could come out big winners, and workers, small wood products firms, dependent businesses, the wood consuming public, and public forest visitors who depend on road access could all lose. Sociologists might refer to this as an "elite settlement." Two powerful interests (large land-owning industries and the environmental community) would win at the expense of everyone else. National forests would be restored to pre-settlement conditions and become sacred sanctuaries enjoyed by wealthy purists, and large timberlands would generate higher profits.

A settlement between these two powerful interests has not been openly advocated by large forest landowners who have actively opposed the shut-down of harvesting on federal lands. The opposite position is taken by leaders in the environmental movement. But political gridlock might be broken if Congress and state legislatures sought to prohibit timber production on most federal lands in exchange for predictable opportunities to develop commodity values on private lands— an offer neither side might be inclined to refuse. This settlement would not result in a desirable or politically stable solution to the forest conflict.

In addition to passing the costs for an elite's environmental preferences along to consumers who will have to pay more for wood products, an elite settlement would inflict enormous losses on dependent industries, workers, families, and rural towns and counties. Moreover, by further empowering the federal government and large corporations, it would destabilize traditions of family, local voluntary associations, local religious associations, and small independent private property owners and businesses. A blend of large capitalism and state-sponsored religion of nature (government ownership of forests in which sacred values prohibit defilement by road-building or logging) is absolutely inconsistent with the American way of life. It also has unflattering parallels with other societies in which elites promoted benefits for a few at the expense of the many or fostered a religion of nature to facilitate control of the masses.

Smaller Businesses Make More Stable Jobs
One of the advantages of the historic mixture of private and public timber supplies was a healthy balance of large corporate and small family wood products businesses. Large businesses can often do

things for the environment that smaller businesses cannot because they are better capitalized. But family businesses appear to provide a more stable employment base. Some of my current research shows that small and medium-sized sawmills and logging contractors, when taken as a whole, make a far more stable employment base than large establishments. This is because small scale entrepreneurial capitalism continues to work well. Also significant is the fact that these smaller sawmills and logging establishments provide a more stable source of employment than non-forest products manufacturing establishments of similar size in Washington and Oregon.

Traditional family operations that do not have to generate a rate of return sufficient to attract Wall Street capital will persist through tough times when the large establishments (generally corporate sawmills) close their doors. Hence, a shift toward greater concentration of production capacity in large establishments is likely to further destabilize employment—especially in rural settlements where the small and medium sized establishments relied on federal timber enterprises.

Property And Control Over Living

Property rights can have profound implications for sustainability because people are more likely to make sacrifices for the future when they feel they have more control over their lives. Studies in agricultural settlements show that local voluntarism and community cooperation is much higher when farm ownership is concentrated in family operations as compared to corporations.[93] What can be done to make people's lives more predictable by giving them greater control over their futures?

European societies learned over 150 years ago that they could not conserve their forests unless they formally established secure rights for local people in uses of both public and private forests.[94] They faced a more difficult situation than we do, since landless peasants displaced from feudal institutions revolted and caused several bloody wars before their needs were addressed. As noted above, this same lesson is being learned over and over again in developing nations—leading to a consensus that bottom-up approaches will do far more for promoting sustainability than use of centralized policing.

Committed To Helping Sustainability

The United States is faced with a far more favorable situation in which rural people still want to be team players and will work to achieve sustainability if given the opportunities to maintain their way of life and economic incentives that make this change possible. A deep reservoir of commitment, cooperation, and resilient institutions can be tapped by using incentives rather than police or moral manipulation.

Look at how the incentives of market niches and enhanced biological productivity motivated the Hatfields to adopt sustainable range management practices. For years, many private forest owners have been managing their lands for biological diversity because they found it profitable to sell hunting rights, and more diverse forests produced more game. Think what would happen if there was an incentive to grow habitat for the northern spotted owl. Private forest owners would jump on the opportunity to reduce the costs of environmental protection and limit their liabilities under the Endangered Species Act. Because some of the good owl habitat is in previously harvested forests, we could immediately begin to build incentive-based recovery plans that would have good chance of stimulating growth of the owl population.[95]

Incentives Can Steer Owners To Sustainability

A wide variety of things could be done to provide incentives. I will briefly mention a few, and leave it to creative economists to continue developing promising approaches: tax credits, special capital gains treatment, publicly-sponsored conservation bonds, and government subsidies for producing public goods such as biological diversity and scenery. But what needs to be emphasized in all these approaches is that private market incentives can be an enormously powerful engine for positive change if it is pointed in the right direction. Americans are ignoring their strengths if they follow along with the radical, puritanical crusade to eliminate the "greed" of the private sector by regulating it with top-down controls or smothering it with moral persuasion.

Guarantees Needed To Rebuild Trust

Re-establishing cooperative relationships with people who rely on federal lands is essential for seeking sustainability. I do not see how this can be done without providing legal guarantees. Legally protected rights are essential because trust is so broken it cannot be repaired without something more substantial than government promises. People are now aware that their lives can be disrupted when agencies such as the Forest Service and Bureau of Land Management are captured by powerful interests—regardless of whether these interests favor timber production or restoring forests to pre-settlement conditions. Average citizens have discovered what sociologists have known for a long time. The "cycling of elites" through leadership positions in public organizations can produce tremendous instability.

Given the susceptibility of federal land management agencies to capture by special interests, I think it is time Americans re-examined whether they want one-third of the nation's land base owned by the government—especially when the environmental preservationist movement seizes control of administrative agencies, forces the government to break long-standing promises and adopt Social Darwinist ethics by abandoning the people it had made dependent upon government resources. What some of the new leaders may eventually learn is that they forfeited the moral authority of the government to administer lands when they failed to act in socially responsible ways.

Ways Of Empowering Local People With Rights

Government ownership of land is proving to be a far greater threat to liberty than many of us had thought. We are learning that individual citizens can only be secure when they are extended legal rights to own or use land or products therefrom. In addition to transferring federal lands to private owners, there are several promising alternatives for establishing rights. Two of these have been tried in other circumstances. Community development corporations have been tried in cities as a means for stimulating self-reliance and attracting the necessary capital for redevelopment. Within counties, port districts or other jurisdictions could be chartered as quasi-public economic entities. These entities could be vested with the authority to raise capital and sponsor investments in new enterprises that would diversify the local economy and localize the circulation of wealth generated from labor and land.

But most importantly, these corporations could contract with federal land management agencies to obtain raw materials for local enterprises or make other uses of public lands. Government agencies would then be obligated to compensate development corporations for breaches of contract if they suddenly found themselves unable to provide materials. Legal obligations would force the government to become more creative in solving its management problems. Government agencies would be forced to behave more like a private company in which decisions to set land aside would have to be weighed very carefully. Such institutional innovation will become an absolute necessity in a world where the demands of a growing population cannot be met without finding ways for increasing production while also protecting biological diversity and environmental health.

Community land trusts are another option for providing predictability by establishing rights. Trust doctrine defends the rights of the beneficiaries and obligates the trustees to serve these interests. Trusts can be used to protect the interests of particular groups or the public. Community trusts could be legally constituted to provide local governments or even a community development corporation in a designated jurisdiction with a claim to rents or other benefits from a tract of forest land owned by the federal government. Or the government's land could be transferred to the trusts, and managed like other trusts. As with community development corporations, community trusts would require the granting of authorities and responsibilities by Congress and state legislatures.

Rights Matched By Responsibilities

Both community development corporations and land trusts would offer the public a chance to specify the responsibilities of local jurisdictions to protect environmental qualities and ecological services enjoyed by the public at large. Hence, both land trusts and development corporations could serve as a midway point between privatization of federal lands and full government ownership. Both would have to comply with all federal and state laws, including the Endangered Species Act, but would enjoy at least some of the security and flexibility of private ownership as well as providing opportunities for citizen control afforded by local community ownership.

Good Places To Start With Empowerment

There are several promising candidate community associations, especially the Association of O & C (Oregon and California) Counties in western Oregon (originating under special legislation to manage railroad grant lands that reverted to federal ownership after the Oregon and California Railroad failed to build the rail system it had promised). Other possibilities are some of the Federal Sustained Yield Units established under the authority of the Cooperative Sustained Yield Act of 1944. These units required the government to manage its timber sales programs to assure wood supplies for local sawmills. If legally protected, the more stable small and medium-sized sawmills and logging operations would benefit from such arrangements, thereby assuring local settlements of a relatively stable employment base and income stream.

Ecological Value Of Local Knowledge

Although secure property rights are essential for living sustainably, knowledge of how to use the land without impairing its long-term productive capacity is also essential. Long-term residents who have managed or worked the land are reservoirs of practical experience. Their knowledge can be tapped to improve management practices and assure protection of essential ecological processes.

For his doctoral dissertation at the University of Washington, Dr. Paul Chandler studied peasants who managed inter-cropping of trees and food crops in southern China. He discovered that farmers knew how to sustain the production of Chinese fir, but that it had taken hundreds of years of trial and error learning. Their "folk knowledge" matched what scientists had learned over several decades of research. For her master's thesis, Catherine Richardson explored the application of this approach to people who were managing land on the Olympic Peninsula in Washington State. Her preliminary findings suggest that the ability of managers to solve ecological problems was related to how long they had worked on the land, and unrelated to level of education or the organization for which they worked.

Untapped Reservoirs Of Knowledge

Such reservoirs of practical knowledge and resilient local institutions are rarely tapped when centralized government agencies consider what lands to use or how to use them. Yet these wise people are often precisely those who know most about how to implement sustainable practices. Empowerment of local people through adopting bottom-up approaches to planning and management would focus this knowledge on the challenges of living sustainably. And the utilization and scientific augmentation of such local knowledge could significantly multiply the effectiveness of research and education expenditures and relieve the government of the need to fund research and development activities that might duplicate what people already know.

Sudden Change Erodes Ecological Knowledge

Precipitous decisions, such as those devastating the rural populations which have relied on federal lands, can contribute to the loss of local knowledge that has accumulated over several generations. In the case of farmers, loggers, and fishermen whose families came from abroad, techniques developed over hundreds of years in Europe are invaluable for developing land use practices that are less disruptive to ecological systems. This traditional knowledge holds more than cultural value; it is a source of information for organic farming, horse logging, and other practices that enable people to work more sustainably with nature.

How Rural Producers Can Re-empower Themselves

Rural producers can bring government back home by focusing their creative energies on finding better approaches to living sustainably with land and telling this story to the American public:

(1) Organize local discussion groups to address questions about the social and environmental responsibilities of the federal government as a landowner.

These discussions should involve a broad group of citizens, including educators, community leaders, clergy persons, environmental advocates, landowners, agency officials, and others.

Questions might focus on issues such as the responsibilities of the federal government to (a) compensate local government when its holdings occupy a large fraction of the potential tax base and local government incurs additional costs associated with uses of the federal lands, (b) assist local workers and business people impacted by decisions to transfer benefits from local people to citizens at large, and (c) help build the capacity for local people to govern their own lives and invent better ways of using the land.

(2) Urge Congress and the Administration to do at home what the United States recommends abroad by legally empowering local people with rights to own or use natural resources.

This can be accomplished by considering a wide variety of options for transferring rights, including fee ownership transfers, easements, leases, or jurisdictional vehicles such as community development corporations and land trusts. Most local benefit will result from restricting transfers to entities that have the maximum likelihood of re-circulating income from land and resources in local jurisdictions. Small businesses, community development corporations, and land trusts seem to offer the most promising opportunities for keeping wealth in rural areas.

(3) Demand federal agencies conducting Environmental Impact Assessments of proposed plans to account for possible losses in traditional local resource management knowledge and expertise.

Social and cultural impact statements could be required to document the presence, value, significance of local knowledge. The risks of losing local knowledge can be compared with the risks of losing species or natural habitats, since both are irreplaceable forms of information. To assure administrative attention, citizens should appeal plans that fail to adequately consider impacts on local knowledge.

(4) Form a local non-governmental organization (NGO) and apply for official recognition by the United Nations.

NGOs have served as useful instruments for empowering of local people in other nations when their governments ignored their needs, undertook oppressive measures, or violated human rights. California Women in Timber has applied to the United Nations for NGO status on the basis of U.S. government policies that have led to the abandonment and material suffering of rural wood products workers. The international forum can be used to build solidarity with rural peoples elsewhere in the world and to publicize deficiencies in U.S. Government policies and actions. International networking has the potential for empowering rural producers throughout the world, especially if they stand by one another in applying pressure on their governments.

(5) Join or organize grass-roots groups and affiliate these groups with others at regional, state, national, and international levels.

Rural political minorities can most effectively gain power by joining together and acting cohesively on issues that matter most. Such groups are likely to be most effective if they embrace rather than fight change, and endorse sustainable development or other themes widely shared in the world community.

(6) Inventory, summarize, and publicize widely all inspirational success stories where rural people adopted sustainable natural resource management practices.

Just as John Kennedy inspired a generation with his book, *Profiles in Courage*, so rural producers can inspire the nation with confidence in the commitment and ability of people to implement sustainable development without government coercion.

13

Telling
Truth

*Science, and more especially scientific progress, are the results not of iso-
lated efforts but of the free competition of thought ... progress depends very
largely on political factors; on political institutions that safeguard freedom of
thought; on democracy.*

Karl Popper
The Poverty of Historicism

    *We are on the threshold of an information technology revolution
that will dramatically transform the way we live and work. New technologies facilitating the exchange of information will make it possible
for people to acquire vast amounts of knowledge that was previously
only accessible to experts or those with the time and resources to visit
libraries and read articles and books. Soon people will be able to sit in
their homes, offices, or retreats and communicate with libraries, data
banks, and other people anywhere in the world where facilities exist. But
it is now more important than ever that we have ways of determining
whether the information we get is trustworthy.*

    *Electronic communication will enable average people to
exchange enormous amounts of information without being face-to-face.*

Technicians at firms such as Microsoft already foreshadow ways of living that many of us will encounter in the future. They "talk" through computer networks, exchange tremendous amounts of technical and scientific information, and work "together" to solve complex problems of designing new technologies. But most interestingly, many of their close "friends" are persons they have never met or even seen in a picture. Yet they "talk" every day by E-mail, and even share deep inner secrets and tell one another about their search for truth—much as I have done in this book. Such "disembodied" communication raises all kinds of questions about the way we will live in the future. It will change where we live, where we work, what we do at our work, and how we create a sense of community. I will discuss how electronic communications may change how we distinguish true and false statements—how we tell the truth.

Living By Reputations

For centuries, most people lived on farms or ran shops. They lived where they worked. Their children grew up watching them work, and by their early teens had acquired many of the skills necessary for following in their parents' footsteps. People were known for what they did for a living, and their status in a community was determined by their job. They identified themselves with their work and community; that was who they were. And, most important for the topic at hand, they determined what was true and false by daily interacting with other people, watching them, gossiping with them, and maintaining consistency in their own behavior so that their personal reputation would remain sound. People who lied, were dishonest, or cheated others were easily discovered, and suffered loss of personal reputation. Remnants of this way of life are still found in many small rural communities, even in countries with largely urbanized societies.

Social Effects Of Industrial Revolution

The industrial revolution brought dramatic change. Work was separated from home and often even from residential communities. Men, and sometimes women, left home early in the morning and did not return until evening. Children were sent to school at a separate location and taught by specialists in education. Education prepared people for specialized jobs that might differ from those of their parents. People were known for more than what they did for a living, and saw themselves in terms of the roles they played at different times of the day or week; father or mother, neighbor, employee, boss, parishioner, student, playmate, or friend. Formal and impersonal means of social control took the place of personal interactions. And new means for establishing truths and falsehoods arose.

Truth Through Science

Science superseded the role of religion as the dominant method for determining what was true and untrue, and established itself as an independent authority. Truth was determined by applying the scientific method. The scientific method turned out to be a tremendously productive way of discovering truths. It gave rise to unprecedented advances in technologies for growing crops, protecting us from disease, synthesizing useful chemicals, and providing us with shelter, clothing, and transportation. These advances were possible because scientists held one another accountable to a rigorous set of rules.

Rules required scientists to focus their attention on testing hypotheses. Hypotheses are guesses about how the world works. The scientific search for truth was most productive when individual scientists were free to study the hypotheses put forward by other scientists and undertook independent research to challenge these propositions. When these hypotheses failed to stand up to challenge, scientists would put forward their own hypotheses and try their best to refute their own guesses about how the world works. Thus, like an economic market that coordinates independent buyers and sellers, the scientific method involved the self-directed coordination of a multitude of independent attempts to challenge hypotheses.

Specialized Elites

The scientific method was so successful that educators, political leaders, industrialists, and others turned increasingly to specialized scientific and professional authorities for guidance on what to believe about the workings of nature and human societies. Authority was concentrated in the hands of professionals such as doctors, dentists, lawyers, foresters, and a wide variety of scientists. These specialists gathered respect and, as social elites, were allocated substantial control over the environment and people's lives. They were trusted to serve the interests of society or individual clients.

Information Age Weakened Elites

The arrival of the information age has dramatically altered the role these specialists play in society. One of the first signs of change was the citizen's or client's interest in participating in professional judgments. As average citizens gained increasing access to knowledge, they began to question the authority of professionals and specialists. They demanded the right to share in the decision-making process. They asserted that professionals did not always know what was best for their clients, and were sometimes wrong about the facts.

As a result, doctors gave patients choices between alternative treatments and asked for "informed consent," public land managers asked citizens to join in "public participation," and attorneys were required to disclose possible conflicts of interest. Specialists were taken off of their pedestals and were no longer protected by their elite status.

From Participation To Information Empowerment

Open access to electronic communications will have an even more profound effect on the role of specialists. Average citizens are already reading specialized journals and scientific reports to empower themselves with information before they accept their doctor's recommendation, trust the judgment of the local forester on the effects of herbicides, or believe the water quality specialist about the probable health effects of minor pollutants. Soon citizens will be able to access all this information from their home or office computer, and, by careful study can learn as much about recent scientific findings or experiences as highly trained specialists. Moreover, computers are increasingly being used to substitute for human judgment when they can do a better job of anticipating what will happen next and how to best respond, including everything from medical prognosis to fire-fighting.

Citizen Empowerment Requires Rules

These new "information super-highways" can empower the average citizen with knowledge and the capacity to search for truth or they can provide opportunities for unprecedented control over people's lives. It all depends on the rules that are established for "driving" these "highways." Hence, we need to watch very closely to assure that those who write the rules for "information traffic" seek to expand and maintain the open exchange of information. The rules need to make it difficult for people to deceive themselves or others about what is actually going on around them. Rules have to motivate people to search for the truth and tell the truth.

But we also need to assure that these information super-highways do not further concentrate wealth and influence in the hands of the "haves" while further isolating the "have-nots." Unequal distribution of access to computer communications now so evident in the cities[96] is also a worry for rural educators. Rural people will be unable to compete effectively in the coming communications revolution unless their children are well educated, have access to the latest technology, and learn how to use it.

Rules Needed To Prevent Elite Control

Moral persuasion poses an especially grave threat to establishing rules that will motivate people to tell the truth. It appears to encourage "spontaneous" and "free" sharing of ideas, but only allows this exchange for purposes of realizing the aspirations of a ruling elite. This was clearly illustrated in the attempt to establish total control over the exchange of information used in the Clinton Administration's forest planning efforts in the Pacific Northwest.

Open exchange of ideas and facts were welcome only so long as they were consistent with the aspirations of the biocentric elite in charge of developing forest plans. A rush of "spontaneous" information exchange took place, but it was limited to producing "options" and "solutions" that fit the elite's preferences. Opportunities for discovering how to maintain wood production while also protecting endangered species were not examined, and those who claimed to have information that could possibly realize these opportunities were excluded from the planning process. As in more totalistic societies, communication was controlled by "gatekeepers" who only allowed "green facts" to enter.

New Parochialism In Electronic Communities

Such limitation on the free exchange of information is the antithesis of the scientific method. Hence, demanding that information exchange adhere to the rules of the scientific method offers a promising antidote for the constricting effects of moral persuasion. Insisting on open scientific debate is obviously a major part of the answer. However, the problems created by electronic communications are far more difficult than can be solved by debate alone.

Electronic communication already permits people to form specialized, insular communities of interest in which they can easily restrict their sources of information to what they want to believe or are comfortable thinking about. The closed thinking of these specialized communities blinds people to other values and ways of life. Thus, urban consumers are cut off from rural resource producers by the information networks in which they participate, not just by geography. People easily delude themselves into thinking that their group has achieved the "truth" and, hence, is doing other people a favor by imposing its "enlightened" view of the world on them. This new parochialism is a product of restricted communication networks, not the geographic isolation that often caused such closed ways of thinking in pre-industrial societies.

Avoiding Electronic Tribalism

Electronic communications can compound such parochialism by making it possible for people to re-establish earlier forms of residential community in which work, residence, and home life are integrated. Information age specialists are already relocating in desirable residential communities, working at home, communicating with clients or supervisors through computer networks linked by satellites, and forming close face-to-face relationships with residential neighbors. Rural communities are often the location for this new lifestyle, especially along the north Pacific Coast and in the inter-mountain region of Montana, Idaho, and Utah.

Yet unrealized is the possibility for people to intentionally cluster together in culturally homogeneous residential communities. Given American tendencies toward Calvinism and Social Darwinism, such social and cultural segmentation has brought with it a wave of socially divisive electronic tribalism. Contemporary radical environmentalists, with their moral condemnation of "other people" (especially resource producers), may simply be a harbinger of a social form that is likely to become increasingly prevalent as we enter the information age. There are signs that such social and cultural segmentation is also taking place along lines of race (growing inter-racial violence), religion (especially mega-churches as full-service communities), and lifestyle (especially counter-cultural expressions based on gender and sexual orientation).

Democratizing Information Through Science

While there is an obvious need for modifying democratic politics to resolve conflicts among culturally segmented groups, my primary concern here is how we are going to discipline the search for truth and govern the expression of claims to truth. The information highways of the future must be governed by rules that make it difficult for each cultural group to reinforce its borders, defend its own version of the truth, or impose its parochial views on others. Advances in scientific methodology provide part of the answer.

Most sciences start with "ruling theories" about how the world works, or "should" work.[97] Facts that might threaten the theory are ignored or discarded. The favored idea is the scientist's intellectual child. An almost paternalistic pride and protective instinct leads scientists to gather supporting facts, assemble allies, and defend the theory against all criticisms. But scientific understanding does not advance very rapidly until scientists begin to translate their ruling theories into hypotheses that can be challenged by facts. This is a difficult transition because it requires scientists to try their best to eliminate their favored intellectual offspring. For this reason, they learn it is easier to advance understanding if they have lots of working hypotheses. Their affections can be spread among a lot of intellectual children with the hope that one or two will survive. At the most advanced stage, scientists set out to efficiently challenge a series of working hypotheses.

Fields such as biochemistry are at this advanced stage of development, while much of the study of forests and natural resource systems, especially forest ecology, is still struggling to emerge from ruling theory. As illustrated by the FEMAT process, spotted owl research, and assessment of rangeland conditions, scientific debates over the natural environment seldom revolve around attempts to challenge working hypotheses. Like early phases in other sciences, the debates are more often ethical, philosophical, and moral, since the objective is to defend a favored view about how the world works. The ruling theory that has dominated natural resources management and planning for most of this century is the idea that nature tends to be stable when undisturbed by humans, and that human activities threaten the "fragility" of ecosystems. Hence, the use of moral persuasion by forest "scientists" is in part an attempt to defend ruling theories about the harmonious workings of undisturbed nature.[98]

Rules Empowering Citizens With Information

Average citizens can help write the rules governing information highways if they understand how scientists are supposed to hold themselves accountable. Citizens cannot be expected to trust information unless its credibility can be established. There need to be standards for licensing people to drive "information highways."

Citizens can assure that scientific information is not distorted by the moral persuasion of social elites, a blind commitment to a ruling theory, or self-delusions stemming from parochialism if they:

(1) Refuse to accept the credibility of scientific claims from government agents, scientists, or private interest groups (both environmentalist and industrial) unless the following conditions are met:

   (a) There is a clear statement of the problem addressed by the scientific claim, and the problem is stated as a question to be answered by facts;

   (b) There is a clear statement of the values or concerns that made this problem important to investigate;

(c) Hypotheses are stated as tentative answers to the question;

(d) Possible weaknesses of the hypotheses are stated in a way that permit them to be rejected;

(e) There is a summary of what was done to attempt rejection of these hypotheses (prove them false);

(f) Conclusions about findings are restricted to tentative answers that survived attempts to reject them;

(g) Other attempts to reject these hypotheses are summarized (replications); and

(h) Critical reviews of the scientific procedures and results are noted and made available (peer review).

(2) Write Congressional representatives to demand that any legislation authorizing the federal government to sponsor or help develop "information highways" contain language giving citizens the right to challenge the credibility of scientific claims if they violate the conditions described above.

(3) Write federal administrative agencies and demand that they develop protocols governing any factual and scientific information they release, and that these protocols incorporate the conditions described above. Refuse to accept the credibility of any information that does not conform to these protocols.

(4) Urge local schools, community colleges, universities, and government extension service offices to develop training courses that empower citizens to distinguish propaganda from defensible factual statements, and to ask questions

that will quickly establish the credibility of factual statements.

5) Appeal to state legislatures to make college and university funding contingent on adoption and enforcement of a code of ethics designed to make faculty members accountable for how they use science or other scholarship to influence political decisions. The protocol mentioned above should be considered as a standard for developing a code of ethics for scientists.

Scientific Attitude Can Restore Trust

The critical, doubting scientific attitude necessary for telling the truth in the electronic age should not be seen as a threat to personal beliefs—even when spiritual or religious commitments are involved. As described in "Finding Peace with Myself," I have recovered ownership of who I am and become more responsible for the environment by adopting the scientific attitude toward myself. My personal search for truth was accelerated by stripping away the false layers laid down when I uncritically adopted someone else's views about who I am, what I should think, and who I should become. Other people may have discovered equally effective ways to disengage the controls put in place by elites who use moral persuasion to control people. But the scientific attitude has been my way of recovering personal freedom, and it has enabled me to restore trust and faith through my own search for truth.

14

Restoring
Liberty

Throughout this book I have tried to demonstrate that people are not angels and cannot rely on ideals or good intentions to live responsibly with one another and their environment. We instead require strong mediating institutions to protect us from the less desirable aspects of ourselves. As individuals, we need the external guidance of strong families, communities, religion, voluntary associations, and democratic governance through a carefully balanced system of legislative, judi-

cial, and executive powers. A government based on laws and the political and spiritual freedom of citizens, not the good intentions of well-meaning people, is essential not just for protecting us from ourselves and from one another, but for protecting the environment as well.

Given human nature, there is precious genius in a government which protects rights to private property, free speech, peaceful assembly, and freedom of religion. We are encouraged to take responsibility ourselves—to "own" all of ourselves. This tells us that as individuals we have both the right and the responsibility to be the best that we can be, given all our warts and blemishes. It also tells us we are accountable for what we do to the environment and what we leave for generations to come.

Private property has been important to American people because it has been fundamental to individual liberty. Ownership of home, land, or business enables people to maintain ownership over themselves. They can be themselves—"own" their selves—when in the privacy of their homes, working their land, or building and maintaining their businesses. As we saw in the story of Chonyi the Magid, property also provides a means for people to leave a tangible part of themselves for the future. This is why the taking of private property is so emotional and politically explosive. To take people's property is to take part of who they are; it is to invade and take part of the self. Yet this is exactly what moral persuasion sets out to do. Moral persuasion is designed to weaken the independent basis of self-identification and moral judgment.

A person cannot become a "new self"—especially one that identifies with a collective such as membership in the egalitarian "community of nature"—unless the "old self" grows weak, dies, or fades away. Taking of private property for purposes of protecting the environment is not just a judicial, legislative, or administrative act. It is also a political act driven by a moralistic crusade to replace the "old self" with a "selfless" concern for the welfare of all living beings. Private property is seen by many as wrong because it protects the autonomy of individuals to behave in "selfish" (homocentric) ways. In the ideal world of a truly communitarian society there would be no private property, because there would be no "private" selves. The "ecosys-

tem" would be owned in common, and a selfless dedication to the environmental security of all beings would be the fundamental moral commandment for living.

Moral persuasion by centralized elites changes the rules of the game. It says that the authority for deciding who we are and what we shall do resides with a few who know best how to live, and that the rest of us have the responsibility to do our best to realize these centralized aspirations. The moral voice, the vision of what is considered to be a right or wrong way to live, shifts to a centralized elite. Hence, it should be no surprise that centralized elites concerned with protecting the environment would elicit the aid of organized religion in their moral crusade to form the "new person."

On October 4, 1993, Vice President Gore assembled religious leaders from across the country to announce the inauguration of the "National Religious Partnership for the Environment," and mark the beginning of "... a new national effort to raise awareness and action on issues of environmental justice and protection ...." Four religious groups were represented: the U.S. Catholic Conference, National Council of Churches in Christ, Consultation on the Environment and Jewish Life, and the Evangelical Environmental network. Dr. Carl Sagan, the charismatic astronomer, joined this meeting on the White House lawn.

In a press release for the meeting at the White House, members of the Partnership claimed to represent more than 100 million Americans when they announced a mobilization campaign. They stated that "Ancient faith communities are hereby resolving to integrate a new world historical challenge throughout all dimensions of religious life," and *"This is not only as a contribution to global justice and sustainability but* an affirmation of what it must mean from now on to be truly and fully religious." *[my emphasis].*

The Partnership was not a spontaneous, grass-roots ecumenical expression of common environmental concern. It all started in January 1990 when a group of politically active scientists got together and wrote and "Open Letter to the Religious Community" to raise awareness of environmental problems. That letter stated that "problems of such magnitude and solutions demanding so broad a perspective must be recognized from the outset as having a religious as well as a scientific dimension."

Their efforts to mobilize congregations on behalf of saving the environment received political support when Senators Al Gore and Tim Wirth were joined by Senators John Heinz and James Jeffords in the creation of the "Joint Appeal by Religion and Science for the Environment." A subsequent meeting between members Congress, scientists, and religious leaders involved the chief executives of eleven national environmental advocacy groups, including the Audubon Society, the Natural Resources Defense Council, the Sierra Club, the Environmental Defense Fund, and the World Resources Institute. At the conclusion of their June 2 - 3, 1991 meeting, the Joint Appeal issued a declaration which stated in part, "We believe a consensus now exists, at the highest level of leadership *across a significant spectrum of religious traditions, that the cause of environmental integrity and justice must occupy* a position of utmost priority for people of faith." *[my emphasis].*

Another Washington, D.C. meeting was held from May 10 through 12, 1992, involving a wide spectrum of religious representatives and leading scientists. In a statement of Joint Appeal resulting from that meeting, scientists asked the "world religious community" to "commit in word and deed, and as boldly as required, to preserve the environment of the earth." Religious leaders responded by affirming the "spirit" of the appeal, and the National Religious Partnership for the Environment was born. Religious leaders representing faiths separated for as long as 2,000 years suddenly "discovered" a community of common concern with one another by succumbing to the climate of fear shared by environmental preservationists. They held hands and knelt before the idol of the environmental apocalypse. The symbolic violence of fracturing the "ecological whole" became their god.

Although not stated in their declaration, the clear implication is that the mission of "saving the environment" rivals or supersedes loving one's neighbor as the "utmost priority for people of faith." A primary concern with faith as the basis of human responsibility was diluted by embracing a biological vision of a "community of nature." Protecting the "ecosystem" became a shared source of security for leaders terror-

ized by the fears of environmental catastrophes. Fear had tempted main-stream religious leaders to turn from their "God of Abraham" and seek environmental security by following the Pied Pipers of apocalyptic violence symbolized by the "fragile web of life." My father taught me this was idolatry by demonstrating that love and faith were stronger than fear.

The Religious Partnership has succeeded in raising $4.5 million from philanthropic foundations with commitments to environmental advocacy. A three-year campaign to "educate" (the Partnership's word) congregants was initiated by Vice President Gore at the October 4 meeting on the White House lawn. Planned actions include "education and action" kits to be distributed to congregations early in 1994, training programs for clergy and lay leadership, legislative updates and testimony on actions, and creation of a "1-800 Green Congregation Hotline" to document grass-roots religious environmental activities.

Penetrating The Wall Between State And Church

The force of moral persuasion is clearly recognizable in this attempt to "educate" congregations, just as it was apparent in the campaign to preserve the "Ancient Forest" in the Pacific Northwest and my own experiences with the environmental movement. I have done my best to find evidence that would tell me that moral persuasion is *not* the best explanation for how people have gone about attempting to save forests or convince citizens to adopt new environmental ethics. But I found it especially hard to deny how a few highly respected scientists, elected officials, and clergy could so easily penetrate the wall separating church and state and use offices of the federal government to promote religious reforms.

I hope I am wrong in drawing this conclusion, and I trust others to take me to task if I am. But if I am right, then moral persuasion is a lot more pervasive and firmly established than I have witnessed in my own experience or in sincere attempts to save a few old-growth forests in the Pacific Northwest. As De Tocqueville predicted and Stephen Carter warned, religion appears to have "accommodated itself to the policies that the state prefers."[99] And, as Carter also warned, we are losing religion as an independent source of "moral understanding without which any majoritarian system can deteriorate into a simple tyranny ...."[100]

Moral Persuasion An Old Friend Of Religion

Accommodation of religion to the apocalyptic fears of radical environmentalists would not be at all unusual. History has taught us that formally organized religion has used moral persuasion as its most effective instrument, and has not hesitated to join hands with powerful political and economic elites when such alliances served its interests. As our Calvinist history has taught us, no institution is more practiced at establishing tension between a corrupt "old self" and a clean "new self" than highly organized religion. All the state or radical environmentalists have done is to adopt and shape to their purposes techniques developed long ago by religious institutions.

Organized religion has been equally masterful at mobilizing people by coordinating the projection of their own unrecognized evil tendencies on heretics and heathen upon whom lethal violence was frequently inflicted. The Crusades, Inquisition, and Christian churches of Nazi Germany are only a few notable examples. There were many individual saints in organized religious groups, but there are almost no organized religions that have consistently behaved in saintly ways—especially when they joined hands with the state or ruling political elites.

Religious Autonomy Basic To America

The genius of the United States Constitution remains its separation of church and state. For as Carter explained so perceptively, it was not the state that needs protection from religion, but religion that needs protection from interference by the state. He emphasizes the need for religious autonomy, by which he means that citizens

> ... should not be beholden to the secular world, that they should exist neither by the forbearance of, or to the bidding of, the society outside of themselves. It means, moreover, that they should be unfettered in preaching resistance to (or support of or indifference toward) the existing order.[101]

Without religious autonomy, there is little to protect us from a few very well-meaning people who delude themselves into thinking they have found the ultimate truth and are justified in monopolizing moral authority.

Violating Religious Autonomy

That is exactly what the "green" clerics are doing. By allowing themselves to be used as instruments of social control by a small group of well-meaning public officials, scientists, and environmental advocates, they are denying people religious autonomy. The Joint Appeal, and all that followed from it, ethically, if not legally, violates the separation of church and state guaranteed by the Constitution. American citizens have a right to practice their religions without having agents of their government take leadership for reshaping the "utmost priority for people of faith" to fit the aspirations of a centralized elite. To claim that one group knows best what the rest of us should believe is a totally unwarranted deviation from American political culture and constitutional principles.

"War" As Means For Monopolizing Moral Authority

The attempts of a few leaders to monopolize moral authority regarding environmental issues is the most serious of the many contemporary uses of moral persuasion. Justifications for taking private property, restricting free speech, and limiting rights of voluntary association could easily follow from the concentration of moral authority in a centralized elite. All that leaders would need to do is to find an effective way of calling upon people's fears of an environmental apocalypse to create a state of excitation necessary for war mobilization.

As we already know from experiences such as the incarceration of Japanese-Americans in World War II, property rights, free speech, and human rights are readily subordinated to the central goal of winning a war. For this reason, few contemporary issues are more important than preventing government officials from in any way attempting to manipulate religious beliefs as a means for promoting generalized fear. There are few greater risks to our freedoms and human dignity than the current attempts to promote a religion of nature by using moral persuasion.

What Citizens Can Do To Protect Freedoms

What can concerned citizens do to protect their freedoms from government officials increasingly inclined to use moral persuasion? The first thing to remember is that our basic political institutions are sound, and offer all the authority needed to protect freedoms when they come under threats. Citizens already hold the powerful instruments they need to protect their freedoms. In addition to legal challenges, they can withdraw the moral support government officials need to maintain legitimacy, work with Congress to withdraw the money the government uses to "educate" the citizens, demand that Congress or the President withdraw officials who violate freedoms, and require their government to abide by both the letter and the spirit of the Constitution. And citizens can get the attention of religious institutions by making their contributions contingent on the replacing obedience to the environmental apocalypse with the worship of God.

Concerned citizens can protect and restore their freedoms by joining together to:

(1) Document and publicize widely all instances in which government officials have interfered with religion, taken private property without fair compensation, or denied free speech or association. Use national newsletters, mass media, and public demonstrations to raise awareness of these threats;

(2) Use the courts to stop government officials from abusing the public trust and violating constitutional principles. Political challenges to the role of elected officials in efforts such as the Joint Appeal should be explored;

(3) Request Congressional investigations and/or hearings to address allegations that government officials have used their offices to interfere with the rights of citizens. The taking of property rights associated with implementation of the Endangered Species Act is an example, as is the role of elected officials in promoting the Joint Appeal and working with the Religious Partnership;

(4) Lobby Congress and the President to terminate funding for government programs which use moral persuasion to "educate" citizens. This might include programs sponsored by the National Advertising Council, The Smithsonian Institution, research projects on topics such as the "spiritual values of forests," and funding for public broadcasting and the arts and humanities;

(5) Develop, maintain, and publish a "freedom for sustainable living" report card to publicize the record of elected officials on environmental issues involving threats to individual freedoms;

(6) Use instruments such as term limitations to help prevent elected officials from becoming dependent on the support and council of powerful elites. Actively support and campaign for candidates who act to protect freedoms and work against the election of candidates who contribute to the erosion of freedoms;

(7) Lobby Congress to reduce the size and discretionary power of federal bureaucracies with authority for regulating activities on private lands. Support the shift of responsibility and authority for environmental regulation from federal to state governments; and

(8) Lobby state legislatures to shift as much authority and responsibility for environmental regulation on private land from state policing agencies to local government agencies and consensus-based stakeholder groups which have reached compromises that balance public interests and private rights.

Closing Comments

Americans face a difficult challenge in addressing the environmental problems they create. Few argue with the need to substantially reduce pollution of air, water, and soil, and to maintain essential ecological functions in streams, rivers, lakes, oceans, and deserts, and forests. Americans want a healthy environment, both for the present and future generations. They disagree on how to achieve this goal.

Respected leaders in government, research, education, media, the clergy, and environmental advocacy groups have increasingly turned to moral persuasion as a means for bringing citizens into conformity with their vision of how we should live with nature. The vast majority of citizens are also concerned with the deteriorating quality of the environment, but have been troubled and confused with the methods used by this elite group. They do not understand moral persuasion and have not learned how to confront it constructively. As a result, messages of fear, guilt, and shame are undermining respect for human dignity, and causing many people to deny the reality of environmental problems.

My purpose in this book has been to affirm the need to take responsibility for environmental problems, but to do so by building from the strengths of American political traditions. When people understand the game of moral persuasion and see how it is being used to monopolize moral authority, they are more likely to distinguish the reality of environmental problems from radical environmentalism's attempts to engineer a new society. I firmly believe we will then discover that the best way to address environmental problems will be to restore and guarantee liberty. Americans are far more likely to take responsibility for solving problems when they remain in control of their lives, are given meaningful incentives, and understand the need to hold themselves accountable for taking corrective actions.

I will bring this book to a close by drawing on the truths I have learned to complete my conversation with the unemployed logger with whom I began. This is how the conversation might have ended had I known then what I know now.

"I can't say for sure what is happening to America, but I can tell you what I think. I think you are among the first to fall in a war over the environment. Your wounds are the pain and suffering of being thrown

out of work and publicly humiliated."

I could sense his anger rising to the surface in search of a target, and continued, "But this war is unlike any war you fought abroad. This is a war of words and emotions—an assault on the hearts and minds of people like you by fear-ridden idealists who want to change who you are and what you think and feel. They have tried to control you by making you feel guilty for cutting trees and shaming you into doing something else for a living—even though they live in wooden houses, use hundreds of products derived from wood, drive cars that pollute the air, eat meat almost every night, and make sure their children have excellent educations and all the advantages in life."

"You're not telling me anything I don't already know," he responded abruptly. "Tell me what we can do to stop it."

"Let me first tell you what won't work," I asserted. "You can't defend yourself in a war of the mind by using violence. If you are honest with yourself, you probably feel like doing some pretty terrible things to strike back. But such outbursts of anger would make your life far worse, because it would surely convince them you deserve to suffer and would remove any sympathy from the public as a whole."

He smiled sheepishly in acknowledgment of how I had read his inner feelings, looked at the ground, and then picked up a stick and studied it. He twisted it in his hand to remove the bark.

"Truth, not violence, is the only way of stopping the war over the environment. And the truth doesn't come out of books or in sophisticated words from the mouths of professors. Truth comes from the heart—from your heart."

He twisted the stick tightly between his fingers to remove the last fragments of bark and rubbed the smooth, golden wood with his finger tips. He looked up, caught my eye in his and issued a challenge, "How much guts do you have, Bob?"

"Enough to match you," I jabbed back with a wink and a grin.

"Do you want to have some fun?" The spark of life returned to his eyes as he broke the stick and dropped it to the ground. " You keep studying those mind-twisters who are tryin' to change us and we'll do our best to carve out a beachhead of freedom for taking this country back one small step at a time. We'll tell our story to all who will listen,

learn new ways of doin' things, take care of the land, and make sure our kids have a future. And who knows, if enough of us do the same thing, we might just help restore America's faith in herself. By God this is gonna' be one hell of a ride watchin' those social engineers break their picks on us. "

Footnotes

Preface

[1] Margaret L. Wheatley. 1994. *Leadership And The New Science: Learning About Organization From An Orderly Universe.* San Francisco: Berrett-Koehler Publications.

Chapter 1 – At War With Ourselves

[2] The description of moral persuasion relies heavily on Charles E. Lindblom. 1977. *Politics and Markets: The World's Political and Economic Systems.* New York: Basic Books, Inc. For an application to and evaluation of moral persuasion to ecological thinking; see John S. Drysek. 1987. *Rational Ecology: Environment and Political Economy.* Oxford and New York: Basil Blackwell.

[3] See especially, Thomas Sowell. 1987. *A Conflict of Visions: Ideological Origins of Political Struggles.* Quill/William Morrow, NY.

[4] Quotations from Mao and Castro were taken from Lindblom, *Politics and Markets*, p. 56.

[5] William James. 1910. "On the Moral Equivalent of War." Reprinted in *The Writings of William James*, ed. John J. McDermott. New York: Modern Library (1967).

[6] For a discussion of how the United States has been almost continually mobilized for one national emergency or another ever since World War I, see Robert Nisbet. 1988. *The Present Age: Progress and Anarchy in Modern America.* New York: Harper and Row.

[7] For a discussion of how perpetual war mobilization undermines liberty, see Friedrich A. Hayek. 1944. *The Road to Serfdom.* Chicago: The University of Chicago.

8 Senator Al Gore. 1992. *Earth In the Balance: Ecology and the Human Spirit.* Boston, New York, London: Houghton Mifflin Company, p. 294.

9 Lindblom, *Politics and Markets*, pp. 277-278.

10 Senator Gore, *Earth in the Balance*, p. 275.

11 Alan Drengson. 1993. "Remembering the Moral and Spiritual Dimensions of Forests," p. 21-23. In Bill Devall, ed., *Clearcut: The Tragedy of Industrial Forestry.* San Francisco: The Foundation for Deep Ecology (published by Sierra Club Books and Earth Island Press).

12 Alexis de Tocqueville, 1969. *Democracy in America.* New York: Doubleday, Anchor Books.

13 Robert N. Bellah, Richard Madsen, William W. Sullivan, Ann Swindler, and Steven M. Tipton. 1985. *Habits of the Heart: Individualism and Commitment in American Life.* Berkeley, Los Angeles, London: University of California Press, p. 271.

14 Quoted from Lindblom, *Politics and Markets*, p. 53.

15 Lynn White, Jr. 1967. "The Historical Roots of Our Ecologic Crisis," *Science* 155 (March 10): 1203-1207.

16 Yi-Fu Tuan. 1968. "Discrepancies Between Environmental Attitudes and Behavior: Examples from Europe and China." *The Canadian Geographer* 12:176-91.

17 See Roderick Frazier Nash. 1969. *The Rights of Nature: A History of Environmental Ethics.* Madison: University of Wisconsin Press.

18 See editorial in the *Seattle Times* (Sunday March 20,1994), "Forests in the Age of the Philosopher-King."

[19] Kai N. Lee. 1993. *Compass and Gyroscope: Integrating Science and Politics for the Environment.* Washington, D.C. and Covelo, Calif.: Island Press., p. 164.

[20] Ibid., p. 165.

Chapter 2 – Obeying Ecological Authority

[21] Stanley Milgram. 1974. *Obedience to Authority.* New York: Harper Torchbooks. I re-encountered Milgram's work through the penetrating insights of Zygmunt Bauman. 1989. *Modernity and the Holocaust.* Ithica, N. Y.: Cornell University Press.

[22] FEMAT (Forest Ecosystem Management Assessment Team). 1993. Forest Ecosystem Management: An Ecological, Economic, and Social Assessment. Portland, Oregon.

[23] Chadwick D. Oliver. 1992. "Achieving and Maintaining Biodiversity and Economic Productivity." *Journal of Forestry* 90(9):20-25

[24] Northwest Forest Resource Council vs. Mike Espy, Secretary of Agriculture. For a discussion of how FEMAT violated the principles of the scientific method, see Robert G. Lee. 1993. A Constructive Critique of the FEMAT Social Assessment. An Independent Paper Prepared for the American Forest and Paper Association, California Forestry Association, and Northwest Forestry Association. Seattle: University of Washington, College of Forest Resources, and Robert G. Lee. 1993. Comments on the DSEIS on Management of Habitat for Late Successional and Old-Growth Forest Related Species Within the Range of the Northern Spotted Owl: with Special Attention to Issues Facing O & C Counties. Prepared for Association of O & C Counties. Seattle: University of Washington, College of Forest Resources.

[25] "Timber-Town Crisis Hits the Children, Too," *Seattle Times* (March 29, 1994).

[26] Ashley L. Schiff. 1962 . *Fire and Water: Scientific Heresy in the U.S. Forest Service.* Cambridge, Mass: Harvard University Press.

[27] See especially, Nash, *The Rights of Nature.*

[28] J. W. Thomas, E.D. Forsman, J.B. Lint, E.C. Meslow, B.R. Noon, and J. Verner. 1990. A Conservation Strategy for the Northern Spotted Owl: A Report of the Interagency Scientific Committee to Address the Conservation of the Northern Spotted Owl. Portland, Oregon: U. S. Department of Agriculture, Forest Service; U.S. Department of Interior, Bureau of Land Management, Fish and Wildlife Service, and National Park Service.

[29] K.N. Johnson, J.F. Franklin, J.W. Thomas, and J. Gordon. 1991. Alternatives for Management of Late-Successional Forests in the Pacific Northwest. Scientific Panel on Late-successional Forest Ecosystems. A Report to the Agriculture and Merchant and Marine Fisheries Committees of the U.S. House of Representatives. Washington, D.C.

[30] J.W. Thomas, M.G. Raphael, R. G. Anthony, E.D. Forsman, A. G. Gunderson, R.S. Holthausen, B. G. Marcot, G. H. Reeves, J.R. Sedell, and M.M. Solis. 1993. Viability Assessments and Management Consideration for Species Associated with Late-Successional and Old-Growth Forests in the Pacific Northwest. Portland, Oregon: U.S. Department of Agriculture, Forest Service.

[31] During post-FEMAT interviews with me, FEMAT scientists who designed Option 9 of the President's report described how they had planned to break the political gridlock by circumventing pluralistic democratic politics and establishing local consensus groups to work in partnership with federal ecosystem managers.

[32] David A. Bella. 1987. "Engineering and the Erosion of Trust." *Journal of Professional Issues in Engineering* 113(2): 117-129, p. 124-125.

[33] Ibid., p. 123.

[34] Ibid., p. 127.

[35] Bauman. *Modernity and the Holocaust.*

Chapter 3 – Beyond Calvin's Ghost

[36] For a discussion of the importance of Calvinism to American culture, especially as it affects the way people view one another see Richard L. Rubenstein. 1983. *The Age of Triage: Fear and Hope in an Overcrowded World.* Boston: Beacon Press. Rubenstein's work is especially important because it alerts Americans to cultural themes that could, under adverse circumstances, lead to genocide of those morally excluded from the community of the elect.

[37] I am indebted to Richard Rubenstein, *The Age of Triage*, for drawing the connection between Calvinism and Social Darwinism.

[38] Rubenstein, *The Age of Triage*, makes a compelling theological interpretation of how people "worship" the laws of nature and the laws of competition.

[39] See especially, Robert Bellah. 1975. *The Broken Covenant: American Civil Religion in Time of Trial.* New York: The Seabury Press.

[40] Rubenstein. *The Age of Triage.*

[41] Published in the March 1990 issue of *P3, The Earth-based Magazine for Kids.* Also see *Captain Planet*, the Saturday morning cartoon for similar "educational" messages.

Chapter 4 – False Dawn
42 The work of the French social psychologist, Serge Moscovici has been especially instructive. See Serge Moscovici. 1985. *The Age of the Crowd.* Cambridge: Cambridge University Press, and Robert M. Farr and Serge Moscovici, eds. 1984. *Social Representations.* Cambridge: Cambridge University Press.

43 Bernard Baruch. 1932. Forward to Charles Mackay. *Memoirs of Extraordinary Popular Delusions and the Madness of Crowds.* Wells, Vermont: L.C. Page & Co.

44 Norman Cohn. 1961. *The Pursuit of the Millennium: Revolutionary Messianism in Medieval and Reformation Europe and its Bearing on Modern Totalitarian Movements.* New York: Harper Torchbooks, p. 74-75.

45 René-Daniel Dubos. 1976. "Symbiosis Between Earth and Humankind." *Science* 193 (August 6): 459-462.

46 See Robert J. Lifton. 1992. "Doubling and the Nazi Doctors." Pp. 218-222 In Jeremiah Abrams and Connie Zweig, eds. *Meeting the Shadow: The Hidden Power of the Dark Side of Human Nature.* Los Angeles: Jeremy P. Tarcher, Inc.

Chapter 5 – Kill The Pig
47 I have drawn the term shadow from Abrams and Zweig, *Meeting the Shadow.*

48 Sam Keen. 1992. "The Enemy Maker." In Abrams and Zweig, *Meeting the Shadow*, p. 201-202.

49 Ibid., p. 201.

50 Rene Girard. 1987. *Things Hidden Since the Foundations of the World.* Stanford: Stanford University Press, p. 155.

[51] Peter Bishop. 1992. "Wilderness as a Victim of Progress," In Abrams and Zweig, *Meeting the Shadow*, p. 155,

[52] Ibid., p. 121.

Chapter 6 – A Forester's Shadow

[53] Franz Heske, *German Forestry*. New Haven: Yale University Press, pp. 180-181.

[54] Ibid., p.114.

[55] Ibid., p. 182.

[56] Robert Jay Lifton. 1983 . *The Broken Connection: On Death and the Continuity of Life*. New York: Basic Books, Inc. ,p. 302.

[57] Heske, *German Forestry*, p. 180-181.

[58] Jan Christiaan Smuts.1967. *Holism and Evolution*. New York: Viking Press, p. 107.

[59] Ibid., p. 345.

[60] p. 272.

[61] Heske, *German Forestry*, p. 28 .

[62] Ibid., p. 42.

[63] Anna Bramwell. 1987. *Ecology in the 20th Century: A History*. New Haven and London: Yale University Press,p. 11.

Chapter 7 – Green Guilt

64 Robert Jay Lifton. 1963. *Thought Reform and the Psychology of Totalism: A Study of "Brainwashing" in China.* New York: W.W. Norton and Company, p. 419.

65 The threat of total social control and the following list of means by which it is established are summarized from Lifton, *Thought Reform.* op. cit., pp. 419-437.

66 Lifton, *Thought Reform, op.cit,* p. 425.

67 Ibid., p. 421.

68 Gregg Easterbrook. 1994. "The Birds: An Environmental Parade." *The New Republic* (March 28); pp. 22-29.

Chapter 8 - Haunted Sanctuary

69 Page Smith. 1990. *Killing the Spirit: Higher Education in America.* New York: Penguin Books, p. 30.

70 Ibid., p. 224 .

71 Yi-Fu Tuan. "Discrepancies Between Environmental Attitudes and Behavior, op. cit." .

72 Robert G. Lee, 1990. "Institutional Stability: A Requisite for Sustainable Forestry," In The Starker Lectures 1990—Sustainable Forestry: Perspectives on the Pacific Northwest. College of Forestry, Oregon State University, Corvallis, Oregon.

73 Robert A. Pois. 1983. *National Socialism and the Religion of Nature.* London and Sydney: Croom Helm.

74 Mircea Eliade. 1974. *The Myth of the Eternal Return, or, Cosmos and History.* Princeton, New Jersey: Princeton University Press.

[75] For a discussion of deep ecology, see Nash, *Rights of Nature.*

[76] These comments were taken from a discussion of her new book: Sallie McFague. 1993. *The Body of God: An Ecological Theology.* Minneapolis: Fortress Press.

Chapter 9 – Talking To Death
[77] My reference to "symbolic murder" is more than rhetoric, since my thinking was deeply influenced by Rene Girard's hypothesis that religions are founded by real or simulated human sacrifice. See Girard, *Things Hidden Since the Foundations of the World*, and Rene Girard. 1977. *Violence and the Sacred.* Baltimore, Maryland: The John Hopkins University Press.

[78] T. S. Elliot, "Little Gidding"

Chapter 10 – Getting Real
[79] These expressions of sustainability were drawn from a talk by Bob Legg of the Temperate Forest Foundation.

[80] Walter Firey. 1963. "Conditions for the Realization of Values Remote in Time," In Edward A. Tirakian, ed., *Sociological Theory, Values and Sociocultural Change: Essays in Honor of Pitirim A. Sorokin.* Glencoe, Illinois: Free Press, p. 150.

[81] Ibid.

[82] See Walter Firey. 1990. "Some Contributions of Sociology to the Study of Natural Resources," In Robert G. Lee, Donald R. Field, and William R. Burch, Jr., eds., *Community and Forestry: Continuities in the Sociology of Natural Resources.* Boulder and San Francisco: Westview Press.

[83] These ideas were most fully developed in Walter Firey. 1960. *Man, Mind and Land: A Theory of Resource Use.* Glencoe, Illinois: Free Press.

[84] The limitations of moral exhortation have been explored systematically in Drysek, *Rational Ecology.*

[85] Elinor Ostrom. 1990. *Governing the Commons: The Evolution of Institutions for Collective Action.* Cambridge: Cambridge University Press.

[86] See for example, IUCN, UNEP, WWF. 1992. *Caring for the Earth: A Guide for Sustainable Living.* Gland, Switzerland.

[87] Erich W. Zimmerman. 1951. *World Resources and Industries* (rev. ed.). New York: Harper & Row, p. 376.

[88] Firey, "Conditions for the Realization of Values Remote in Time," op. cit., p. 157.

Chapter 11 – Celebrating Sally's Heart

[89] Jonathan Kusel and Louise Fortmann. 1991. Well-Being in Forest-Dependent Communities. Sacramento, California: California Department of Forestry and Fire Protection, Forest and Rangeland Resources Assessment Program (FRAP), p. 14.

[90] Christopher Lash. 1988. "The Communitarian Critique of Liberalism," In Charles H. Reynolds and Ralph V. Norman, eds., *Community in America: The Challenge of Habits of the Heart.* Berkeley, Los Angeles, and London: University of California Press, p. 182.

Chapter 12 – Bringing Government Back Home

91 Institute of Forest Resources, University of Washington. 1990. "Three-State Impact of Spotted Owl Conservation and other Timber Harvest Reductions: A Cooperative Evaluation of the Economic and Social Impacts," Institute of Forest Resources, Contribution No. 69, September.

92 This is the cumulative effect of job losses attributed to all causes, including forest plans, owl conservation plans, and the current Clinton Administration forest plan.

93 Walter R. Goldschmidt. 1978. *As You Sow: Three Studies in the Social Consequences of Agribusiness.* Montclair, N.J.: Allanheld, Osmun.

94 See Bernard E. Fernow. 1913. *History of Forestry.* Toronto: University Press, and Washington, D.C.: American Forestry Association.

95 Bruce Lippke. 1992. "Meeting The Need For Environmental Protection While Satisfying The Global Demand For Wood And Other Raw Materials." In James Boyer, ed. In Proceedings of the EPRS conference on wood products demand and the environment. Vancouver B.C.

Chapter 13 – Telling Truth

96 "The Information Gap." *Newsweek*, March 21, 1994.

97 For further reading on the development of science see two exceptionally valuable articles: T.C. Chamberlain. 1890. "The Method of Multiple Working Hypotheses." Reprinted in *Science* 148: 754-759 (1965), and J. R. Platt. 1964. "Strong Inference." *Science* 146: 347-353.

[98] For an extended discussion of these competing views of nature from the viewpoint of an ecologist, see Daniel B. Botkin. 1990. *Discordant Harmonies: A New Ecology for the Twenty-First Century.* New York and Oxford: Oxford University Press. Also see the insightful work of Chadwick Oliver, a silviculturist at the University of Washington.

Chapter 14 – Restoring Liberty
[99] Stephan L. Carter. 1993. *The Culture of Disbelief: How American Law and Politics Trivialize Religion.* New York: Basic Books, p.38.

[100] Ibid., p. 36.

[101] Ibid., pp. 36-37.